STP 1112

Pesticide Formulations and Application Systems: 11th Volume

Loren E. Bode and David G. Chasin, editors

ASTM Publication Code Number (PCN)
04-011120-48

ASTM
1916 Race Street
Philadelphia, PA 19103

Library of Congress

PCN: 04-011120-48
ISBN: 0-8031-1414-1
ISSN: 1040-1695

Peer Review Policy

Each paper published in this volume was evaluated by three peer reviewers. The authors addressed all of the reviewers' comments to the satisfaction of both the technical editor(s) and the ASTM Committee on Publications.

The quality of the papers in this publication reflects not only the obvious efforts of the authors and the technical editor(s), but also the work of these peer reviewers. The ASTM Committee on Publications acknowledges with appreciation their dedication and contribution to time and effort on behalf of ASTM.

Printed in Ann Arbor, MI
April 1992

Foreword

This publication, *Pesticide Formulations and Application Systems: 11th Volume,* contains papers presented at the symposium of the same name, held in San Antonio, TX on 14–15 Nov. 1991. The symposium was sponsored by ASTM Committee E-35 on Pesticides and its Subcommittee, E35.22 on Pesticide Formulations and Application Systems. Loren E. Bode of the University of Illinois at Urbana and David G. Chasin of ICI Specialty Chemicals in Wilmington, DE, presided as symposium co-chairmen and are editors of the resulting publication.

Contents

APPLICATION TECHNIQUES AND PEST CONTROL

Overview

The 11th Symposium on Pesticide Formulations and Application Systems began the second decade of increased interaction between pesticide formulators, application scientists, and regulatory personnel. For increased efficiency and safety of pesticide use, more knowledge of the technical and regulatory aspects of formulations and application parameters must be obtained and shared among the disciplines involved with pesticides. The purpose of the symposium was to:

- Provide a forum for exchange of ideas and data among chemists and engineers working to improve the efficiency of pesticide use.
- Provide a data base to support ASTM Committee E–35 in development of guides and standards.
- Serve as a guide to Subcommittee E35.22 members in their future efforts to address the issues related to the use of pesticides.

This volume, in addition to previous symposia proceedings, adds significantly to the available resources on the important subject of pesticides. Topics in this STP include the technical aspects of pesticide application and formulation research, including equipment and concepts contributing to the effective and responsible use of pesticides. Safety aspects of pesticides are included as an integral part of pesticide development and use. Direction and suggestions for the development of standards were made available from material presented at the symposium.

The 22 papers in this STP are organized into three sections. The first section, *Safety and Environmental Impacts*, includes seven papers regarding humans, food, environmental issues, and presents new technology to insure that our safe supply of food is maintained. The second section, *Characteristics of Formulations and Adjuvants*, includes research regarding pesticide formulation and adjuvant technology. Seven papers regarding pesticide application comprise the final section, *Application Techniques and Pest Control*. This section includes research on new application systems and evaluations of pest control.

Safety and Environmental Impacts

Pesticide effects on the environment is a major factor in effective pesticide use. The papers in this section discuss some of the safety aspects of using pesticides to produce our food supply. The paper by Cummings compares the proposed California food safety initiatives and discusses the potential effects on formulation chemists and inert ingredient suppliers. Kogan and Gieseke's paper presents a computer model that provides quantitative information on human exposure to airborne particles in closed environments. Point source contamination of groundwater from loading and mixing sites can be prevented by use of a CARBO-FLO water treatment developed by Ohio State University. The treatment process uses a simple flocculation and filtration system that is described in the paper by Hall, Downer, and Chapple.

Controlled release granular formulations have the potential of reducing the amount of pesticide applied. Papers by Shasha and Wing present state of the art research regarding encapsulated herbicides. Stein describes a method to monitor release of pesticides from granules into the soil, and Meyers showed that the change in release rate of chemicals in response to temperature is unique to microcapsules made from Intelimer polymers.

In addition to the formal papers, D. Lindsay and B. Omilinsky, EPA subcontractor Formulogics personnel, were invited to address the complex issues on the proposed rules regarding container management that is contained in FIFRA 88. The discussion provided a very interactive conclusion to the 11th symposium that was of value to the domestic as well as many international attendees who actively provided input regarding the rules being developed by the EPA.

Characteristics of Formulations and Adjuvants

Narayanan and Chaudhuri present a working model for emulsifiable concentrate formulations that explains the generality for the concentrates and high stability of emulsions observed on dilution with water. Duckworth and Cearnal present the results of a study to determine the effect of carrier temperature on the emulsification or dispersion of pesticide formulations. They suggest that the current ASTM standard be revised to include water temperatures that depend on the end use conditions of the pesticide. Becher shows that for testing compatibility in the laboratory, carriers must be representative of the type actually used in the field. Tann, Berger, and Berger demonstrated the application of dynamic surface tension to the study of adjuvants and emulsion stability.

Application Techniques and Pest Control

Wright described the new Expedite Pesticide Applicator System that is a combination of ready to use pesticide formulations, closed system packaging, and state-of-the-art, low-volume, controlled-droplet spray delivery. Ozkan outlined a comprehensive and complete set of guidelines which may help revise ASTM Standard E 641 that deals with measuring the wear rate of nozzle tips.

Biological results from pesticide application were presented by Chambers, Prasad, Manthey, and Riley. Chambers paper shows that mineral oil as a carrier for low volume applications was superior to vegetable oils in weed control efficacy. Prasad found that some adjuvants enhance the effectiveness of glyphosate sprays without damaging the crop species and discussed the relation of droplet size to phytotoxicity. Manthey reported on the relationship between phytotoxicity of a postemergence herbicide and the physical properties of surfactants. Riley's paper provides a detailed assessment of spray deposition and efficacy from insecticides, when aerially applied to stands of fir for control of the western spruce budworm.

These papers confirm that the objectives of the symposium were met. This STP (in conjunction with previous symposia STPs, provides a database of information regarding pesticide formulations and application systems that will guide ASTM Subcommittee E35.22 members in the development of necessary standards.

Loren E. Bode, editor
University of Illinois,
Urbana, IL 61801

Safety and Environmental Impacts

GARY L. CUMMINGS

THE CALIFORNIA FOOD SAFETY INITIATIVES: THE IMPACT ON FORMULATORS

REFERENCE: Cummings, G. L., "The California Food Safety Initiatives: The Impact on Formulations," *Pesticide Formulations and Application Systems: 11th Volume*, ASTM STP 1112, Loren E. Bode and David G. Chasin, Eds., American Society for Testing and Materials, Philadelphia, 1992.

ABSTRACT: The November, 1990 California ballot contained two initiatives which would impact the agricultural chemical industry. The most noted initiative, Proposition 128 popularly called the Big Green or Hayden/Van de Kamp initiative, addressed several environmentally popular themes, including food safety. The most significant effect on agrichem companies would come from the outright banning of certain chemicals, independent of concentration or risk, in food production. Products would be banned based on toxic active ingredients, metabolites of actives, impurities in actives, degrades of actives, inerts, metabolites of inerts, and impurities in inerts. Production Agriculture had sponsored a second, alternate initiative as a more responsible approach to food safety, Proposition 135. This paper will compare both initiatives, will discuss the initiative strategies and will discuss the potential effects on formulation chemists and inert suppliers.

KEYWORDS: Environmental Protection Act of 1990, Pesticide Enforcement Act, inert ingredients.

BACKGROUND

Many states have an initiative process. Californians have the process and have been very active in using it to achieve change. The increasing use of initiatives results from voter frustration with the inability, unwillingness or refusal of the legislature to seriously face contemporary issues. Californians have been very proactive as illustrated by a tax initiative (Proposition 13), a toxic chemical initiative (Proposition 65) and with many other lesser known initiatives. This fall, two initiatives will be on the ballot which will have major effects on the agrichemical industry.

The first, and potentially most significant, is Proposition 128, the Environmental Protection Act of 1990 [1]; commonly referred to as "Big Green", "Hayden Initiative" or "Hayden/Van de Kamp".

Gary L. Cummings is Supervisor, Formulations at Valent U.S.A. Corporation, 1333 North California Blvd., Walnut Creek, CA 94596.

This initiative has wide support and deserves close attention, whether or not passed by the voters.

The second initiative, Proposition 135 (Consumer Pesticide Enforcement Act for Food, Water and Worker Safety [2]), was developed by production agriculture as a reaction to the Environmental Protection Act. This act also would have significant impact in the agrichemical industry. However, this act is much more consistent with the current regulatory climate. The California initiative process provided an opportunity for consumers, the food industry and many worker groups to counteract the food safety extremes in the Environmental Protection Act, yet strengthen food safety programs in a meaningful manner. Where provisions of two successful initiatives overlap, the provisions of the initiative with the most votes prevail.

This paper does not intend to discuss all aspects of the two Acts or to discuss the politics and economics. After a general description of the features of the Acts, the paper will focus on the sections that are likely to impact us, as formulators.

OVERVIEW OF ENVIRONMENTAL PROTECTION ACT OF 1990

This Act has been very skillfully designed to appeal to most voters by including several issues of concern to Californians. Included are such environmentally sensitive issues as depletion of the ozone layer, protection of redwood trees, restriction of off-shore drilling, water quality improvement and food safety. Passage had been expected due to the perceived widespread voter appeal.

Careful analysis of the food safety sections leads to the following observations:

Proposition 128

- Imposes arbitrary, inflexible standards;

- Moves regulatory authority from the Department of Food and Agriculture (CDFA) to the Department of Health Services (DHS);

- Eliminates consideration of benefits as well as risks;

- Eliminates scientific judgment from certain decisions;

- Heavily impacts inert ingredients;

- Bans all products that are reproductive toxins and Group A or B carcinogens, including metabolites, degradates, impurities, inerts, inert metabolites and inert degradates;

- Bans Group C carcinogens unless registrants demonstrate safety. The timetables for rebuttal are likely to be unworkable. Many products would be banned because of administrative inactivity, not science. Metabolites, degradates, impurities, inerts are included;

- Sets many inflexible standards for worker safety;

- Sets tougher analytical standards;

- Bans any foodstuffs shipped into or through California which contain residues of banned active ingredients or inerts;

OVERVIEW OF PESTICIDE ENFORCEMENT ACT

This Act also has many features which are of interest to the voters. The single most important feature of Proposition 135 is that science and cost/risk remain as policy features in administering the food and worker safety program. Salient observations of this Act are high-lighted as follow:

Proposition 135

- Utilizes science and science advisory boards;

- Focuses on active ingredients and EPA Inerts of Toxicological Concern (List 1), not on degradates, impurities, metabolites;

- Funds more research into alternatives for pesticides;

- Continues to administer food safety through CDFA;

- Improves safety in transportation of agricultural commodities;

- Funds research and alternatives to medfly spraying;

- Increases state testing of raw and processed foods;

ENVIRONMENTAL PROTECTION ACT - DEFINITIONS

We have briefly examined the general strategies and main features of the two Acts. The next step is to evaluate Proposition 128, the Environmental Protection Act, in more depth. The definitions of several terms are important and provide insight into the comprehensive approach of the Act. These definitions are taken directly from the text (Article 5, 26914).

"Pesticide" or "pesticide chemical" means any substance which alone, in chemical combination, or in formulations with one or more substances, is an "economic poison" as defined by Section 12753 of the Food and Agriculture Code or a pesticide as defined in Section 2(u) of the Federal Insecticide, Fungicide and Rodenticide Act, but including the active ingredient, metabolites, contaminants, degradation product or inert ingredients and which is used in the production, storage or transportation of any food.

"Active ingredient" means a pesticide, excluding its inert ingredients, but including its metabolites, contaminants and degradation products.

"Inert ingredient" means an ingredient that is not active, as defined in Section 2 (m) of the Federal Insecticide, Fungicide and Rodenticide Act and including any contaminant therein or any substance which is the result of metabolism or other degradation of the inert ingredient.

"Contaminant" means a constituent of a registered pesticide which is unavoidably produced during the manufacture of the active ingredient.

"Degradation product" means the result of the biotransformation or breakdown of the parent compound by food processing or environmental factors including but not limited to air, sunlight or water.

"Metabolite" means the result of biotransformation or breakdown of the parent compound by a living organism.

ENVIRONMENTAL PROTECTION ACT - SECTIONS AFFECTING FORMULATORS

Certain features of Proposition 128 will impact formulators directly or indirectly.

BANNING OF PESTICIDES

All active ingredients known to cause cancer or reproductive harm, registered for use on food or having a tolerance, shall be banned by January 1, 1996. No new registrations or tolerances are allowed. No pesticides with missing data or inadequate data shall be registered for food use. (Article 1, 26901).

Any active ingredient is known to cause cancer or to cause reproductive harm if it is characterized by EPA as a Group A or B carcinogen or is on the California Proposition 65 Carcinogen List.

The following factors should be noted:

- There are no risk assessment or de minimus references.

- The active ingredient definition includes metabolites, contaminants and degradation products.

- Trace materials would be included.

- Missing data, tests in progress, etc. would preclude registration of new products.

REVIEW OF PESTICIDES

Producers can petition for review of "high hazard pesticides" which are defined as EPA Class C carcinogens. Petitions must be filed by November 7, 1994 and the DHS must evaluate the petition and data within one year. If the DHS does not respond favorably within one year, the pesticide shall be considered to be known to cause cancer. The criteria for assessment are the criteria of a Group B carcinogen (Article 2, 26903).

The following factors should be noted:

- The time frames are unrealistic.

- Products may be canceled due to DHS's inability to complete reviews within one year. Products may be canceled by default, not science.

- The time frames may not allow development of data to meet the assessment criteria. This would particularly impact metabolites, contaminants or degradates.

BANNING OF PESTICIDE INERT INGREDIENTS

No food use pesticide containing an inert ingredient known to cause cancer or reproductive harm may be registered or granted a tolerance. The time table for cancellation is two years (November 7, 1992). DHS shall not permit the use of any inert known to cause cancer or reproductive harm in the formulation of a pesticide (Article 2, 26904).

EPA Group A or B carcinogens or Proposition 65 listed inerts are banned, Registrants may petition to have Group C carcinogens approved. If the inert is determined not to be a carcinogen, it must still be evaluated to determine that it poses no other risk.

The following factors should be noted:

- No exceptions are given for trace impurities.

- The definition of inerts includes metabolites and degradates.

- The review time tables are unrealistic. Little data may already exist for inerts, ingredients of inerts, metabolites or degradates.

- Inerts may be banned by default, not toxicity.

- California will be breaking new ground in administering inerts.

- "Other risks" is not defined.

PESTICIDE ENFORCEMENT ACT - SECTIONS AFFECTING FORMULATORS

Proposition 135 uses existing definitions within the California Food and Agricultural Code or FIFRA. The term "inert ingredient" is specifically defined as "an ingredient" in a pesticide which is not an active ingredient as defined in the Federal Insecticide, Fungicide and Rodenticide Act (7 U.S.C., Sec. 136 (a)).

The Director of CDFA may require any registrant that has registered a pesticide which contains an inert ingredient which has been listed by EPA as an Inert of Toxological Concern (List 1), or any other inert ingredient which has been deemed by the director to be of significant toxological concern, to submit appropriate and relevant acute toxicity, chronic toxicity and residue data to the department for the inert ingredient in question.

The following comments are appropriate:

- The administration of inerts appears to follow EPA guidelines.

- Risk and science are included in the assessments.

- Impurities, metabolites and degradates are not specifically covered, although the Act allows CDFA to assess and regulate inerts of unspecified toxicological concern.

GENERAL IMPACT ON FORMULATORS

The Pesticide Enforcement Act primarily would appear to strengthen CDFA's administration of inerts, without adding significant requirements beyond those imposed by EPA. However, the Environmental Protection Act would have a profound impact on formulators and suppliers, and many new or expanded demands on formulators can be predicted from passage of Proposition 128.

- Many currently registered products will be banned.

- Formulators are frequently responsible for product degradation studies. The need to elucidate degradation mechanisms and to identify degradates will expand.

- More information will be required on the identity and fate of impurities used by formulators.

- Inert suppliers will need to be educated. Formulator's new responsibilities with respect to inerts, such as impurity or metabolite identification, will be beyond the knowledge, experience, capability or financial interest of many suppliers.

- Great care will be necessary in selection of brands or suppliers of inerts. We will need to work closely with many suppliers to learn for ourselves or help them learn more about their products.

- Many more back-up or alternate formulations will be needed. We will not be able to predict which inerts will become objects of concern, and we will not always be able to predict the outcome of regulatory reviews of inerts.

- Formulation considerations will become more important in business decisions. More emphasis on data development programs, strategies or task forces will be necessary to support specific inert usage.

- Uncertainty will remain over what "other health risks" means.

- Many inerts will no longer be useful options for formulators.

- There will be considerable chaos as agrichemical producers and inert suppliers begin to translate the ramifications of the Act into specific programs.

SPECIFIC IMPACTS ON FORMULATORS

Certain key ingredients are potentially impacted by the Environmental Protection Act. The following list identifies key inerts or impurities of concern. This list results from a relatively superficial survey of commonly used inert ingredients. Many other ingredients will be affected due to contaminants, degradates or metabolites whose potential presence may be indicated by literature searches, more in-depth discussions with suppliers or in-house data. Proposition 65 lists the following substances as chemicals known to the State to cause cancer. These substances are potentially associated with agricultural chemical inert ingredients:

Acetaldehyde, benzene, ethylene thiourea, formaldehyde, silica, lead, dichloromethane, methylene chloride, methylene oxide, ethylene oxide, heavy metals, ethyl alcohol, soots, tars, mineral oils, propylene oxide, 1,4-dioxane.

Certain inerts have particular significance to formulators, since they have wide spread usage in pesticide formulations. The following comments are very general but can be illustrative of problems that will become apparent in inerts evaluations.

Emulsifiers

Most agricultural chemical emulsifiers contain one or more ethoxylates. When you examine the MSDSs for many emulsifiers, you frequently find a reference to the possible presence of ethylene oxide. It is difficult to foresee continued use of one of these emulsifiers under Proposition 128. Fortunately, ethoxylators should be able to eliminate free ethylene oxide and remove the reference from their MSDSs.

Solvents

Tank trucks may carry gasoline or solvents, at times without thorough clean-out in the changeover. Benzene, from previous gasoline loads, may cause contamination of solvents. Fortunately this is a solvable problem.

Clays

The potential presence and characterization of silica (crystalline of respirable size) is a difficult issue for clay producers. Failure to resolve this issue may have a major impact on our ability to formulate granular, dust or wettable formulations. Heavy metal impurities may have an impact also.

SUMMARY

Californians will be voting on two initiatives. The effect on the agricultural chemical industry and pesticide formulation process will be major if the Environmental Protection Act passes, and gains more votes than the Pesticide Enforcement Act. The impact will be much less dramatic if the Pesticide Enforcement Act passes, and gains more votes then the Environmental Protection Act.

We must face a broader issue also. Food safety is a major focus of the environmental activism of the 90s. Inerts, impurities, degradates and metabolites are being drawn into the food safety debate. Decisions as to selection, source of supply, processing, contamination, degradation, and overall research on inerts will play an increasing important role in agricultural chemical business.

REFERENCES

[1]   Environmental Protection Act of 1990 (Proposition 128). California Voter Initiative.
[2]   Consumer Pesticide Enforcement Act for Food, Water, and Worker Safety (Proposition 135). California Voter Initiative.

Vladimir Kogan and James A. Gieseke

MODELING HUMAN EXPOSURE TO AIRBORNE
PESTICIDES IN CLOSED ENVIRONMENTS

REFERENCE: Kogan, V. and Gieseke, J. A., "Modeling Human
Exposure to Airborne Pesticides in Closed Environments,"
Pesticide Formulations and Application Systems: 11th
Volume, ASTM STP 1112, Loren E. Bode and David G. Chasin,
Eds., American Society for Testing and Materials,
Philadelphia, 1992.

ABSTRACT: A computer program was developed for the assessment of
human exposure to airborne pesticides in the work place, such as
a greenhouse or a chemical production plant. The program
considers various aspects of aerosol physics including accounting
for sources, transport, deposition, and evolution of the size
distribution of aerosol materials. In addition, it addresses
phase transition phenomena and photolysis. It evaluates level of
exposure to airborne particles based on arbitrary schedule of
worker activities. Results of model calculations using the
computer code show the importance of source strength, droplet
size, air flow patterns, applicator activities and location, and
atmospheric conditions.

KEYWORDS: atmospheric dispersion, aerosol science, worker
exposure, mathematical model, computer code, airborne pesticides

INTRODUCTION

Agricultural chemicals, including pesticides, play a notable
technological role in the modern food industry. They are also used to
protect our health. Consequently, development, production, and the
use of pesticides have become a necessary part of a variety of
occupational activities. Paradoxically, pesticides are toxic
substances and their ecological impact has well known side effects.
It requires a scientifically sound approach to dealing with pesticides
in order to assure public safety and gain environmental acceptance.

As a result of toxicological considerations, any potential harm
to human health from exposure to pesticides raises a legitimate public
concern. Combined with other factors, it influences negatively the
public's attitude toward the agrochemical industry. In order to
assure occupational safety of workers dealing with pesticides and to
contribute to a more positive and better deserved perception of
agricultural chemicals, an a priori evaluation of the level of an
individual's exposure in the workplace is needed.

Vladimir Kogan is a research scientist at Battelle Memorial
Institute, 505 King Avenue, Columbus, OH 43201; Dr. James A. Gieseke
is a department manager at Battelle Memorial Institute.

Assessment of risks associated with the exposure of personnel to toxic airborne chemicals has become an important health and environmental issue. Exposure can occur in closed environments, such as in a greenhouse, a chemical production plant, a farmer's residence, or a cab or cockpit on application machinery such as an airplane or truck. People can be exposed to airborne pesticides as a result of their indoor use, indoor emissions from process equipment, or leaks to the indoors from the outside. Traditionally, sampling and monitoring of airborne chemicals have been used to determine levels of human exposure. In addition to this approach, which relies on measuring levels of actual exposure, mathematical modeling has proved to be an accurate and inexpensive means for evaluating the potential for adverse effects that may result from dermal or inhalation exposure to toxic aerosols. This analytic approach is appealing in many cases where some measures should be taken before-hand in order to reduce the possibility for excessive exposure.

In the past, aerosol modeling in closed environments has been a subject of scientific endeavor for a wide variety of applications. As a result, a number of computer codes were developed and used in technical areas with divergent interests ranging from nuclear safety [1] to indoor air quality [2]. The main purpose of undertaking this study was to apply to the agrochemicals industry a convenient analytical approach for specifically addressing the safety aspects associated with atmospheric dispersion of pesticides in a microenvironment where excessive concentrations of pesticide particles can be a health concern. With some modifications, this approach can be extended to include the outdoor exposures as well. The PEST computer program was developed under this task. It is founded on broad experience accumulated in various areas of modeling indoor distribution and atmospheric dispersion of particulate matter. It is a mechanistic model and is based almost entirely on fundamental laws of aerosol mechanics rather than on application specific correlations. The term aerosol is defined here as any air suspension of either liquid or solid particles.

There are many potential applications of the computer program described here. It delivers its main product as the exposure dose, in both dermal and inhalation terms, of a person as determined by local environment or concentrations of airborne pesticides and on the schedule of his activities. In addition, it generates information that can be used to address some general aspects of the environmental fate of pesticides, their application efficiency and losses. It can be also used to design or evaluate the effectiveness of safety provisions in a plant where emissions of concentrated chemicals can occur during production or handling operations.

MODEL GENERAL FORMULATION

In general terms, the intent of the PEST program is to predict the atmospheric concentration and behavior of a pesticide after it has been dispersed in air and until it is settled out or vented from the microenvironment, and to evaluate the exposure level for a worker involved in the pesticide application process, or in general, a person who is present in this environment.

Intuitively, a person working in the vicinity of a pesticide source is exposed to higher concentrations of aerosol particles than a person at some distance from the source. This is because of dilutions which occur as the released materials mix into increasingly larger volumes and because of deposition from the air. This point is stressed in the code using an exposure volume concept. According to this concept, the person is located inside of a hypothetical exposure volume which can communicate with the rest of facility, or the main room, by means of convective flows. The rate of air exchange between the volumes depends upon ventilation arrangements within the facility, and is also influenced by an arbitrary flow generated by a local cooling fan, by air-assisted spraying, or alternatively is modified directly to accommodate results of experimental observations. There may be several exposure volumes in the room, but only one of them could be occupied at a time. This unique approach allows sufficient flexibility in postulating a problem using nonhomogeneous distribution of airborne droplets throughout the room and an arbitrary schedule of worker activities. Since air exchange rates between volumes can be adjusted by the PEST user, an exposure volume can be viewed as another actual room adjacent to the main facility.

Atmospheric dispersion of a pesticide (e.g., an application spray, a leak, or emission) occurs in the hypothetical volumes as well. This allows the kinetics of atmospheric dispersion to be addressed and provides an additional dimension to the microenvironment under consideration. There are no spatio-temporal correlations or limitations between locating the worker and activating the source of pesticide dispersion within any of the hypothetical volumes.

In order to accommodate a common situation where ventilation flow is developed in a predominant direction, the facility can also be considered as a string or series of zones. Each zone represents a fraction of the total facility volume (e.g., a room or part of a room) and is in itself a well-mixed volume. More specifically, the room is subdivided into a number of equal zones by imaginary walls across the flow. There may be exposure and dispersion volumes within each of the zones formed in this manner. The number of zones is specified by the user and can total up to ZMAX according to the dimensions of PEST arrays:

$$ZMAX = INT(24/(RVMAX+1)),\qquad(1)$$

where   ZMAX = maximum number of zones, and
         RVMAX = greatest number of exposure and dispersion
                 volumes in one zone.

In this formula, number 24 is the maximum number of all the compartments addressable by the code, and INT means that integer part of the quotient is being taken.

Since the total number of zones can be sufficiently large, a well-mixed environment is assumed in each zone. Consequently, a well-mixed facility with no strongly directed flow will consist of just one zone.

Fig. 1 illustrates the PEST approach to modeling human exposure, where a pesticide applicator is exposed to some level of the aerosolized chemical in a greenhouse. This facility is viewed as a single well-mixed room, and the hypothetical dispersion and exposure volumes are shown as shaded areas.

FIG. 1 -- Illustration of PEST modeling approach.

As a mechanistic model, PEST employs detailed representation of aerosol parameters, addresses appropriate physico-chemical phenomena of dispersion in air, models general flow patterns in the indoor atmosphere, and allows input of time dependent variables. Brief descriptions of some of the PEST concepts follow.

Aerosol Representation. Two levels of numerical resolution are used in PEST for representing airborne particles. First, continuous distribution of particle sizes is shaped using discrete size channels, where total number of particles within each channel is conserved but a single particle size, v^k, is used to represent them. Particle volume, v, is used as a measure of size, and v^k is the geometric mean of the left and right boundaries of the k-channel.

The second resolution relates to aerosol composition. Currently, PEST keeps track of at least three airborne components. Two of them are an active ingredient and a product of its photolysis. The third component is an inert ingredient or a solvent. According to our approach, particles within each size channel are allowed an arbitrary composition that may vary in time and be different for each size channel. However, the aerosol composition is assumed to be the same for all particles of the same size. This composition information is updated every time step as an average for each size channel.

Source Term. The rate of pesticide atmospheric dispersion can be an arbitrary function of time, including zero, and is specified for each of the dispersion volumes. For convenience, initial droplet size distribution is assumed to be log-normal, so that only the mean size, v_g, and the geometric standard deviation, σ_g, are needed to describe, for example, a spray dispersion. This distribution takes the following form:

$$f(v) = \frac{1}{\sqrt{2\pi}\, v\, 3\ln \sigma_g} \exp\left(-\frac{\ln^2(v/v_g)}{2(3 \ln \sigma_g)^2}\right), \tag{2}$$

where σ_g is given for the distribution of droplet diameters.

Flow Modeling. As mentioned above, flow patterns in the main facility are addressed in two dimensions. One dimension, in the direction of wind, or the room ventilation flow, is resolved using a string of zones, which allows one to model both well-mixed and directed-flow cases. Although uniform environmental conditions are assumed in each zone, the second dimension addresses local non-homogeneity surrounding a spray, or an emission source, and is resolved using the hypothetical dispersion and exposure volumes.

Aerosol Deposition on Surfaces. The removal of pesticide droplets, or particles, from air suspension is characterized by the removal flux expressed as the number of particles leaving the environment per unit time and per unit removal area (such as a unit deposition area or a unit cross-sectional area of the outlet flow). A convenient approach to representing the removal flux, ϕ, is to use 'deposition velocity', V_d, defining the flux as $\phi = V_d * C$, where C is the uniform aerosol concentration in the microenvironment. Then, in a well-mixed compartment, the rate of concentration decay due to some ith mechanism with removal velocity V_d^i is expressed as

$$\frac{dC}{dt} = -R^i(v,t)C, \tag{3}$$

$R^i(v,t) = V^i_d(v,t)*A^i/V$; where index i symbolizes a particular removal mechanism due to which particles with size v deposit on surfaces at V_d deposition velocity. A^i and V are the surface area available for deposition due to mechanism i and the volume of the compartment, respectively. V_d is taken as the steady state, or terminal velocity defined as $V_d = B(v)F(v,t)$ where $B(v)$ is the particle mobility, and $F(v,t)$ is the force applied to the particle.

Removal of airborne droplets by sedimentation and diffusion is currently considered by PEST. Other deposition mechanisms, such as those caused by turbulent flow conditions and electrostatic effects, can be readily added to the computational network due to the modular structure of the PEST program. Detailed information on aerosol deposition phenomena is readily available [3].

Droplet Evaporation. Evaporation of a volatile component from a particular pesticide formulation is modeled taking into consideration both the heat and mass transfer rates at the droplet-air interface, and accounts for the surface curvature effects as well [4]:

$$\frac{dv}{dt} = 4\pi r \frac{S - e^{X/r}}{B + Ge^{X/r}} \tag{4}$$

where r is the droplet radius, S is the saturation ratio defined as the ratio of ambient vapor pressure over the saturation vapor pressure, and B, G, and X are functions of a number of physico-chemical characteristics of both the gaseous and liquid phases.

This expression is used in PEST in combination with an arbitrary rate function, which is introduced in order to account for the formulation-specific adsorption forces acting between different components of the pesticide formulation.

According to Eq. 4, in a supersaturated atmosphere, when $S > 1$, there exists a critical droplet radius, r_{crit}, such that condensation occurs only on larger droplets, while the smaller droplets continue to evaporate, resulting in a redistribution of aerosol parameters. This phase transition effect of surface curvature is important in several respects ranging from designing optimal spray parameters to recommending formulation components, and also appears to play a role in evaluating personal exposure. Since fine mists evaporate almost instantly, saturation conditions can be expected in closed environments such as a green house. In addition to this, aerosol cooling due to evaporation of a high-vapor-pressure component, such as a solvent, may result in supersaturation conditions at the droplet surface for a less volatile component, such as an active ingredient, resulting in more efficient and ecologically accepted application and lower drift losses of the chemical. Some experimental observations of this process have been noted [5].

Photolysis. Photochemical decomposition of airborne pesticides is important to consider because definite amounts of finely disseminated chemicals remain suspended in air for a long time, losing both their plant protection capacity as well as toxicological potential. Photolysis is currently approximated in PEST using the first-order kinetics:

$$\frac{dC}{dt} = -kC \tag{5}$$

where k is the rate constant. An arbitrary function of ambient lighting can be incorporated in order to evaluate the rate constant.

Model Equations. According to the PEST approach a general differential equation for modeling atmospheric dispersion of pesticides in closed environments can be written as:

$$\frac{dC(v,1,x,y,t)}{dt} = -V_x \frac{\partial C}{\partial X} - V_y \frac{\partial C}{\partial y} - \sum_i R^i C + \tag{6}$$

$$\frac{S}{V} + \int_v^{v_{max}} H(v)\ \gamma(v)\ C(v)\ dv - H(1-\gamma)C - kC$$

where $C(v,1,x,y,t)$ is the local number concentration of aerosol particles with size v and composition 1 in a two-dimensional space $\{x,y\}$ at time t, S is the rate of pesticide dispersion, H is the droplet evaporation rate, and γ is a droplet redistribution parameter due to the phase transition processes. The x and y dimensions are introduced to resolve possible variability in aerosol concentration in the direction of ventilation flow, as addressed with the main room zones, and within the zones as addressed with the dispersion/exposure volumes, respectively. V_x and V_y are the respective air velocity components in these dimensions.

Applicator Exposure. Evaluation of personal exposure in PEST is addressed in terms of information on the person's activity. Information of this type includes his location at any time of the problem, his exposed areas (both total and horizontal), and breathing rate. This information is provided by the code user.

The dermal exposure level is calculated as the amount of pesticide, in terms of each component, deposited on the worker's open surface areas. This is done in a mechanistic fashion using aerosol deposition modeling as outlined above.

The inhalation exposure to the pesticide can be evaluated using either of the two different models employed in the PEST code. Both these models use the predicted airborne concentration in different compartments in order to obtain either total or inhaled aerosol retention.

The retention of inhaled particles occurs in different parts of the respiratory tract including the extrathoracic airways, tracheo-bronchial airways, and the alveolar spaces. The total inhalation exposure based on combining these regional depositions is employed in one of the PEST approaches. For simplicity, this is done on a semi-empirical basis by interpolating theoretical results of Xu and Yu [6]. It generates a vector of retention efficiencies, η, for particles in the range from 0.01 μm to 10 μm in diameter and provides total retention efficiency for an adult male respiratory tract using two piecewise cubic fits calculated with the method of least squares:

$$\eta = 0.888 - 9.817\ d + 40.642\ d^2 - 48.318\ d^3$$

for $0.01\ \mu m < d < 0.5\ \mu m$, and $\qquad\qquad\qquad\qquad\qquad\qquad$ (7)

$$\eta = 0.051 + 0.070\ d + 0.024\ d^2 - 0.002\ d^3$$

for $0.5\ \mu m < d < 10\ \mu m$, where d is the particle diameter in microns, and η is the respirable fraction of these particles.

An alternative solution also used in PEST is the American Conference of Government and Industrial Hygienists (ACGIH) definition of respirable particulate mass:

$$\eta = 1 - \int_0^\infty \frac{1}{\sqrt{2\pi}\ d\ \ln 1.5}\ \exp\left(-\frac{\ln^2(d/3.5)}{2\ \ln^2 1.5}\right) d(d),\qquad (8)$$

where d is the particle diameter in microns.

In either of the two cases, the inhaled total particle concentration is analyzed for retention in terms of the fraction of particles in each size range multiplied by the fraction retained in each size range.

PEST SAMPLE APPLICATION

There are many complex phenomena that can affect dissemination of agricultural chemicals in the atmosphere. The degree to which a particular phenomenon affects the overall pesticide application process depends on various factors. Therefore, an attempt to accommodate a long list of these phenomena in a single computer program can overcomplicate it and make the computational procedure inefficient. Instead, construction of the program in a modular fashion, where the mathematical models of various processes can be rearranged as desired, has appeared to be an attractive alternative. This approach has been adopted for developing the PEST program. Hence, some of the factors, such as direction of discharge from a spray nozzle or volatilization from treated surfaces, are not currently considered by PEST, but can be readily incorporated into its network as needed. In this particular work, for illustrative purposes, our interests were concentrated on main roles that the flow conditions and the air humidity can play in affecting the level of personal exposure in a greenhouse during pesticide application for a half-an-hour period of time.

For simplicity, our model pesticide formulation is assumed here to consist of a 10% water suspension of an active ingredient which has a low vapor pressure (assumed 1.9E-7 torr at 20°C), is practically insoluble in water (<1 ppm), and has a negligible photolysis rate. An applicator sprays this formulation at 0.1 l/min for the 30-min period of time in a 1000 m^3 greenhouse under normal atmospheric conditions. The green house is ventilated at 2 air exchanges per hour (16.6667 m^3/min). Surface areas of the applicator that are exposed to ambient air are equivalent to his head and two hands, and there is no pesticide penetration through his clothing. His breathing rate is 20 l/min.

The droplet size distribution of the spray has a mass median diameter of 58 μm and a geometric standard deviation of 1.7. Fig. 2 shows both the number and the mass distributions of the spray droplets, which are calculated for the diameter range from 0.1 μm to 1000.0 μm using the log-normal distribution law.

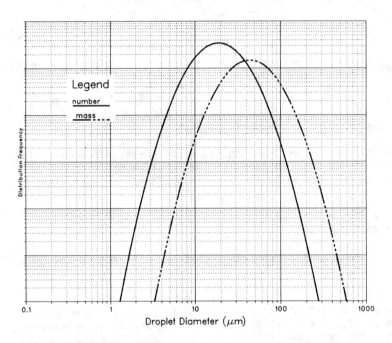

FIG. 2 -- Spray droplet size distribution for PEST model calculations.

Several scenarios of pesticide application in a greenhouse could be envisioned. In order to illustrate capabilities of the PEST program, we have considered three situations, which are schematically shown in Fig. 3. The first two scenarios are similar in that in both cases, there are well-mixed atmospheric conditions considered in the main room. However, in the first case, the applicator is always exposed to the same conditions as in the main room, which is typical for the application when the worker walks away from the sprayed area, while in the second case, the applicator walks forward, i.e., into the sprayed zone. It has been assumed that there is 30 seconds delay before the worker enters this exposure volume, where he stays for another 30 seconds, and then moves on. This means that on the average, he is continuously exposed to that concentration of pesticide droplets that would evolve in a release volume in about 45 seconds after the spray had been relocated after 30 seconds of application.

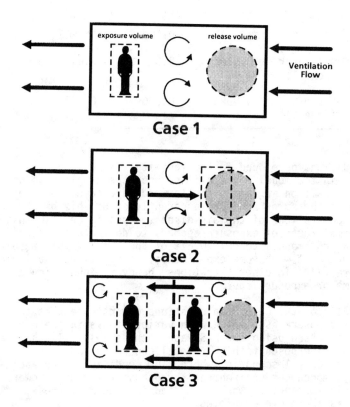

FIG. 3 -- Illustration of postulated cases for PEST model calculations.

The third case considered here was set up to illustrate a situation where ventilation arrangements provide directed-flow conditions inside the facility. In this case, two zones are introduced to facilitate aerosol deposition from air as it flows in the direction of the ventilation stream. There are two workers in the green house, each located in a different zone, and the applicator's location is shown in the same zone with the release volume.

Table 1 summarizes results of PEST calculations made for the three scenarios described.

As the table shows, levels of both dermal and inhalation exposure can be affected by the work pattern of the applicator and by

TABLE 1 -- PEST predictions of personal exposure.

Case No.		Level of Personal Exposure, g		
		Dermal	Inhalation	Total
1		0.008	0.015	0.023
2		15.100	0.102	15.202
3	Applicator (Zone 1)	0.014	0.026	0.040
	Worker (Zone 2)	0.002	0.006	0.008

the flow conditions developed in the facility. With such a fine spray as used in the model calculations, when the droplets do not immediately deposit on surfaces, the forward advancement of the applicator into the spray zone can apparently expose him or her to very high concentrations of the air-dispersed pesticide. In this case, the dermal route of exposure appears to dominate the respiratory route, which amounts to only 0.67% of the total. This result is in general agreement with some experimental observations available in the literature [7]. In other situations, however, inhalation can represent a major exposure route.

The results of the Case 3 calculations demonstrate that the potential for exposure associated with various tasks in the work place can be assessed using the PEST approach as well. In this case, the applicator is exposed to higher local concentrations when compared to the Case 1 scenario, where flow conditions were adequate to provide a homogeneous environment throughout the room. Lower exposure levels for another worker located in Zone 2 of these calculations is therefore predicted.

Obviously, atmospheric conditions are also important when considering air dispersions of volatile compounds. In closed compartments, air humidity will increase due to evaporation of small droplets. Spraying or misting is actually used in some applications for achieving supersaturated conditions. Under these conditions, as mentioned above, vapor condensation on larger airborne particles and surfaces can occur. Fig. 4 illustrates this situation by displaying relative condensation rates of droplets as a function of saturation ratio, S, and droplet diameter. The relative condensation rate is defined in this work as $(dv/dt)/v$, where v is the droplet volume, and t is the time, so that negative values of the rate quantify the evaporation of droplets.

As a result of droplet evaporation/condensation phenomena, a size redistribution of airborne pesticide dispersions takes place. For example, under the postulated conditions of our model calculations as formulated above, Fig. 5 shows the droplet mass distribution in a well-mixed room (Case 1) which would evolve in the greenhouse as

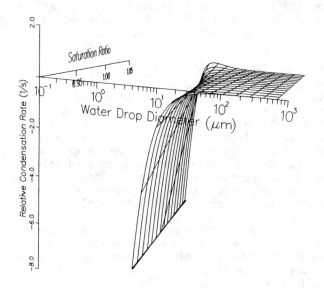

FIG. 4 -- Vapor condensation on water droplets.

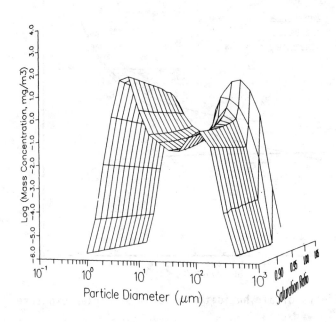

FIG. 5 -- Mass concentration of airborne pesticide droplets in model compartment.

a function of relative humidity after a 30-min pesticide application. This two-dimensional dependence is represented as a surface,
so that a particular droplet size distribution can be viewed as a
slice through the surface, made for any selected value of the
saturation ratio.

There is a well pronounced trough in this surface, which
separates two distribution modes. These two modes evolve in time as
a result of a complex of dynamic processes, including the phase
transition phenomena. In particular, under lower air humidity, these
modes are sustained, to a large extent, due to the evaporative
formation of the fines and their higher persistence in air suspension, and due to the continuous source of the larger spray droplets,
respectively. As the humidity increases, evaporation of droplets
slows and, under supersaturated conditions, vapor condensation occurs
on larger particles. In turn, this results in further development of
the coarse mode (note the logarithmic concentration scale).

The size redistribution of airborne droplets caused by phase
transition of a volatile component affects worker exposure to an
active pesticide ingredient as well. This is shown in Fig. 6 where
the exposure predictions are now displayed for the same model
calculations described above.

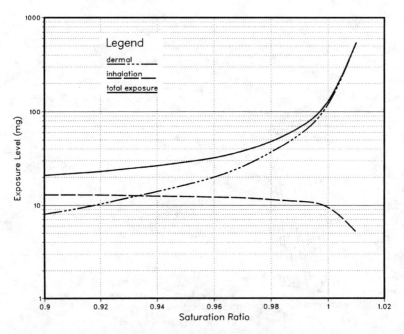

FIG. 6 -- Effects of air humidity on level of worker exposure.

As this figure shows, the dermal route of worker exposure
exemplifies higher risk with higher relative humidity, especially in
a supersaturated atmosphere, while the respiratory route of exposure

diminishes. This could have been expected because of the higher deposition rates of larger droplets, and since the respiratory fraction of airborne pesticides decreases as the droplet size distribution deforms due to higher humidity.

In conclusion, a mechanistic computer program named PEST has been developed to model the dynamic behavior of airborne chemicals and to evaluate their potential effects on workers in their work places. The level of worker exposure to air-dispersed pesticides is calculated using an arbitrary schedule of worker activities and based on existing air flow patterns and atmospheric conditions. Results of PEST sample calculations illustrate that the assessment of worker exposure to airborne chemicals in the work place can be effectively predicted using this analytic approach. Both natural and controllable conditions can be evaluated for importance in affecting exposure.

REFERENCES

[1] Gieseke, J. A., Cybulskis, P., Jordan, H., Lee, K. W., Schumacher, P. M., Curtis, L. A., Wooton, R. O., Quayle, S. F., and Kogan, V., "Source Term Code Package (Mod 1), A User's Guide", NUREG/CR-4587, BMI-2138, report to the U.S. Nuclear Regulatory Commission, 1986.

[2] Nazaroff, W. W. and Cass, G. R., "Mathematical Modeling of Indoor Aerosol Dynamics", Environ. Sci. Technol., Vol. 23, No. 2, 1989.

[3] Fuchs, N. A., The Mechanics of Aerosols, Dover Publications, New York, 1989.

[4] Mason, B. J., The Physics of Clouds, Oxford at the Clarendon Press, 1957.

[5] Seaver, M., Galloway, A., and Manuccia, T. J., "Water Condensation onto an Evaporating Drop of 1-Butanol", Aerosol Sci. Technol., Vol. 12, 1190, pp. 741-744.

[6] Xu, G. B. and Yu, C. P., "Effects of Age on Deposition of Inhaled Aerosols in the Human Lung", Aerosol Sci. Technol., Vol. 5, 1986, pp. 349-357.

[7] Pesticide Asessment Guidelines, Subdivision U, Applicator Exposure Monitoring, U.S. Environmental Protection Agency, National Technical Information Service (PB87-133286), 1986.

Franklin R. Hall, Roger A. Downer, and Andrew C. Chapple

EVALUATION OF THE CARBO-FLO® PESTICIDE WASTE MANAGEMENT SYSTEM.

REFERENCE: Hall, F. R., Downer, R. A., and Chapple, A. C., "Evaluation of the Carbo-Flo Pesticide Waste Management System," Pesticide Formulations and Application Systems: 11th Volume, ASTM STP 1112, Loren E. Bode and David G. Chasin, Eds., American Society for Testing and Materials, Philadelphia, 1992.

ABSTRACT: The CARBO-FLO® water effluent treatment process uses a simple flocculation and filtration system which requires no specialized personnel for its effective use. Treatment is carried out using pre-weighed/pre-packed materials designed to treat 1000 liters of diluted spray liquor. The flocculation and settlement process takes a minimum of 1 hour. Filtration through one coarse gravel and two fine carbon filters takes approximately 2 hours.

Water containing atrazine, alachlor and permethrin was processed through the specially designed SENTINEL® plant. These pesticides were chosen as they are widely used and represent a range of chemical groups. Trials using both single products and a three way mix were performed. Analysis of the treated water at different points in the system was carried out using gas chromatography. All treatments resulted in >99.9% of the pesticide being removed from the end water. Water thus treated can, at present, be recycled for subsequent use as rinse water or may be discharged into a soakaway. All solid waste from the process can be incinerated. The carbon filters can be regenerated. Some possible user problems with the flocculation process were encountered and investigated. Adequate dilution of the pesticide waste liquor was found to be essential for the process to be completely efficient.

KEYWORDS: Pesticide waste management, Permethrin, Atrazine, Alachlor.

Professor Hall is Head of the Laboratory for Pest Control Application Technology (LPCAT), The Ohio State University, Wooster OH 44691. Mr Downer is a post-doctoral research associate and Mr Chapple is a graduate research associate.

The presence of pesticides in the environment is being given much attention by world society. Recent surveys by a number of state and federal agencies have revealed the presence of an array of pesticides in groundwater and well water [1,2]. The 1988 Amendments to the Federal Insecticide, Fungicide and Rodenticide Act (FIFRA-88) proposes to revise the regulations on storage, disposal, transportation, and recall of pesticides and pesticide containers [3]. This is likely to entail greater control of the disposal of pesticide rinsate. Currently, in many cases the lack of proper containment at the pesticide mixing/loading facilities has resulted in the increased potential for point source contamination of soil and groundwater surrounding these sites. Where these facilities are utilized by custom operators, aerial and other commercial applicators, the potential for contamination of surrounding ground becomes more serious as rinsing operations become more frequent. Prevention of point source pollution is clearly the most cost effective strategy. Innovative new technologies combined with long term management practices will ultimately determine the quality of the environment near agricultural pesticide mixing/loading sites. The research described was intended to assess the efficiency of a new pesticide rinsate treatment process designed to safely and economically remove potentially hazardous residues from spray water rinsates.

MATERIALS AND METHODS

A SENTINEL° water effluent treatment plant was delivered to the Ohio Agricultural Research and Development Center (OARDC) in April 1989 from ICI Agrochemicals/E.Allman & Co. Ltd, UK. The construction and assembly were as per manufacturers instructions [4]. The system consists of a 1000 liter tank, pump, flocculation chamber, collection tray (for flocculated materials), one gravel filter and two carbon filters (see Figure 1).

Priming of the system is carried out by half filling the main tank with clean water and then opening the liquid take-off valve, with the outlet valve at filter '3' closed. Water flowing into the filter elements forces air out of the air bleed valves and charges the filters, at which point the outlet valve at filter '3' was opened and the tank drained. Initially the water was colored with carbon fines but cleared quickly. While the tank was draining the system was checked for leaks and the flow rate checked. This was found to be within the manufacturers specifications of 4-6 liters per minute [4].

For initial evaluation of the systems efficiency in removing certain pesticides, permethrin (Ambush° 2E ICI Americas Inc.), alachlor (Lasso° EC Monsanto Company) and atrazine (AAtrex° 4L Ciba-Geigy Corporation) were evaluated as single products and as a three-way tank mix combination. Simulated spray tank waste liquor (500 l) was made up in the main tank by adding the pesticide/s to the water while running the agitator paddle. The order of addition of the pesticides for the three way mixture was 1. AAtrex, 2. Lasso, 3. Ambush. The resultant rate of dilution was intended to simulate a field rate mixture diluted with water as should be the case when returning from routine spray operations and washing out a sprayer. When the tank contents were suitably mixed, a sample of the tank mixture (approx. 100 ml) was removed for later analysis.

The system is designed to treat 1000 liters of <u>diluted</u> spray effluent: concentration of the active ingredients in the effluent must not exceed 0.5%. In this assessment (for convenience) 500 liters was treated for each experiment. Therefore the four pre-packed Carbo-flo® materials, (labeled 1,2,3 and 4), were divided, as necessary, by weight.

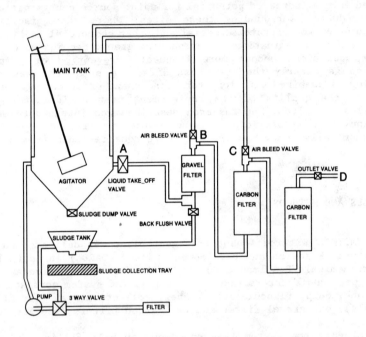

FIG. 1 -- Schematic of the SENTINEL® system.

With the agitator running, the first Carbo-flo® material was sprinkled onto the surface of the effluent. When this was completed, material 2 was added followed by material 3 and the agitation continued for 15 minutes. Flocculation was seen to begin after the addition of the second material. The effluent was seen to have a pinkish color. This was due to a dye deliberately included to enable the user to monitor the activity of the carbon filters. (When the liquid passing between filters 1 and 2 during filtration is seen to be colored it is an indication that filter 1 needs replacing immediately). After 15 minutes agitation Carbo-flo material 4 was slowly added and agitation continued for a minimum of 1 minute. The agitator was turned off and the contents of the tank allowed to settle.

The time allowed for settlement to take place is at the discretion of the operator but must be at least 45 minutes. It is important to allow the sludge to settle below the level of the liquid take-off valve before initiating the filtration process. Under normal working conditions the flocculation and sedimentation process can be carried out either late morning, eg, pre lunch-break or late afternoon, allowing settlement overnight.

The filtration process was started by opening the liquid take-off valve and closing the filter '3' valve to release any air. When bubbles had stopped rising in the air bleed valve tubes, the filter '3' valve was opened.

Once the filtration process had been activated, samples of treated permethrin water were taken at 15 minute intervals from the points shown on the diagram. Point D, the outlet hose, was sampled first. Then points A, B and C in order. This was achieved by closing the liquid take-off valve and disconnecting the hose. Then by carefully opening the valve, 100 ml (approx.) of liquid could be removed. The valve was then closed. Hoses were sequentially disconnected at points B and C and samples taken. The liquid take-off valve was then re-opened and the system allowed to run for a further 15 minutes. Sampling was then repeated. Since the flow rate through the system diminished as the volume in the main tank decreased a total of 11 samples were taken from points A,B and D but only 10 from point C.

For atrazine and alachlor, samples of treated water were taken at intervals from points A, B and D as shown on the diagram.

All samples were analysed for pesticide content by using standard gas chromatography techniques. Extraction of pesticide from the aqueous samples was achieved as follows:-

A 100 ml aliquot of aqueous sample was taken and placed into a separatory funnel with 60 ml of methylene chloride and 10 ml saturated sodium chloride solution. The sample was shaken 100 times and the phases allowed to separate. The methylene chloride layer (bottom) was collected in a 500 ml round bottom flask after filtering through a bed of sodium sulfate pre-wet with methylene chloride. The extraction was repeated twice with 60 ml methylene chloride and the methylene chloride phases were collected in the 500 ml round bottom flask. The sodium sulphate was washed with 15 ml methylene chloride and added to the flask. The methylene chloride was evaporated just to dryness on a Roto-Vap in a water bath at 35 °C. The pesticide was recovered in 10 ml acetone.

The gas chromatograph used for permethrin analysis was a Varian Gas Chromatograph Model 3600 with a Model 8035 Autosampler (1 ul injection).

Column: 6 meter glass; 2 mm Internal diameter.
Packing: 1) SP 2330, 68% cyanopropyl (1%)
         2) Supercoport 80/100 pesticide grade
Carrier gas: High purity (99.9995%) nitrogen
Flow: 30 ml/min
Detector: Electron capture Ni63 (8mCi)

Temperatures: Oven: 215 °C, Detector: 300 °C, Injector: 275 °C

For atrazine and alachlor analysis, the same method was used but with the following modifications:

Column: 6 meter glass; 2 mm Internal diameter.
Packing: DC 200 10% on "Q".

Carrier gas: Helium.
Flow: 20 ml/min.
Detector: TSD.

Temperatures: Oven: 200 °C
                 Detector: 270 °C
                 Injector: 230 °C

RESULTS

The quantity of permethrin (ppm) detected in samples of effluent over the two and a half hour sampling period is shown in Table 1. Initial concentrate contained 237.5 ppm and the detection limit was 0.01 ppm.

The quantity of atrazine (ppm) detected in samples of effluent over the two hour sampling period, is shown in Table 2. The initial concentrate contained 5100 ppm and the detection limit was 0.004 ppm.

The quantity of alachlor (ppm) detected in samples of effluent over the two hour sampling period is shown in Table 3. The initial concentrate contained 795 ppm and the detection limit was 0.0004 ppm.

TABLE 1 -- Effluent treatment efficacy:- Permethrin concentration in ppm.

Time from	Sample points			
start (min)	A	B	C	D
0	125	-	-	-
15	10	0.01	ND	ND
30	0.27	0.01	ND	ND
45	0.58	0.03	ND	ND
60	0.04	ND	ND	ND
75	0.13	ND	ND	ND
90	0.06	ND	ND	ND
105	0.12	ND	ND	ND
120	0.02	ND	ND	ND
135	0.05	ND	ND	ND
150	0.09	ND	-	ND

ND = Not detectable.

TABLE 2 -- Effluent treatment efficacy:- Atrazine concentration in ppm.

| Time from | | Sample points | |
start (min)	A	B	D
0	32.5	-	-
30	25.7	29.8	0.0059
60	26.5	28.9	0.0073
90	25.6	29.8	0.0050
120	23.9	25.9	0.0044

TABLE 3 -- Effluent treatment efficacy:- Alachlor concentration in ppm

| Time from | | Sample points | |
start (min)	A	B	D
30	48.7	38.2	<0.0004
60	61.4	61.1	<0.0004
90	43.5	60.4	0.0048
120	43.3	62.4	<0.0004

TABLE 4 -- Effluent treatment efficacy:- Concentration (ppm) of permethrin, atrazine and alachlor in a 3-way mixture.

Time from start (min)	Permethrin	Atrazine	Alachlor
30	ND	0.084	ND
60	ND	ND	ND
90	ND	ND	ND
120	ND	0.109	ND

The quantity of permethrin, atrazine and alachlor detected in samples of mixed effluent over the two hour sampling period is shown in Table 4. The initial concentrations were 237.5 ppm, 5100 ppm and 795 ppm respectively and the detection limits were as for the single product analysis.

DISCUSSION

When treating water containing atrazine, it was noticed that flocculation and settlement was considerably reduced. The immediate effect of this was that the first carbon filter became blocked and the flow through the system eventually stopped. This filter was replaced.

The problems experienced with the flocculation process appear not to have affected the final outcome of the treatment. Data generated in 1987 by ADAS Research and Development Services [5] shows that some pesticides such as pirimicarb, mecoprop, 2,4-D and paraquat are similarly unaffected by the flocculation process (Table 5), but are removed by the carbon filters.

TABLE 5 -- Pesticide concentrations (ppm) found in treated water [after 5].

Pesticide	Original Concentration	After flocculation	After filtration
Demeton-S-methyl	250	<0.5	<0.02
gamma-HCH	530	17	<0.02
Pirimicarb	180	170	<0.02
Propiconazole	100	19	<0.02
Cypermethrin	25	1	<0.02
2,4-D	61	62	0.01
Mecoprop	119	128	0.01
Paraquat	580	198	<0.5

The failure of the flocculation process during the processing of atrazine was studied in the laboratory, and it was shown that the problem was the concentration of active ingredient in the original mixture. The original mixture contained 0.54% active ingredient, 0.04% greater than the recommended concentration.

Four concentrations of active ingredient were used for this test:-

1. 0.27%, equvalent to half the recommended field rate.

2. 0.5%, max a.i recommended by ICI.

3. 0.54%, equivalent to the recommended field rate.

4. 0.71%, To simulate an extra 2 liters product added to the SENTINEL° in error.

When the appropriate proportions of the Carbo-flo° chemicals were added to the above solutions the following observations were made:-

Solution 1: settled out within half an hour of adding chemicals, and the liquid became a very pale orange.

Solution 2: also settled out, but the liquid became pink and more or less transparent.

Solution 3: also settled out, but the liquid was a lighter pink than solution 2 and was not transparent.

Solution 4: did not settle out, and the whole solution remained a vivid pink.

The conclusion drawn from these observations was that the concentration of active ingredient/s in the original tank mixture must be below 0.5% and that even a small increase in this percentage could result in an unacceptable reduction in flocculation efficiency and subsequent reduction in working life of the carbon filters. Exploration of the efficiency of the system against an array of other pesticides is planned, eg; those with high patterns of use, those with high potential to leach into groundwater and those with extremely high efficiency (low AI/ha)] on fruits, vegetables and field crops. The disposal of the solid waste in the form of sludge is currently being investigated by ICI Agrochemicals plc in the UK. The findings of this investigation will be reported soon.

CONCLUSIONS

1. The Carbo-flo° treatment of spray waste removes >99.9% of pesticide from the water. In the case of alachlor this is equal to or below drinking water standards (EPA).

2. The concentration of pesticide in the main tank must not exceed 0.5%.

3. Formulation and AI type do make a difference to the ease of flocculation and hence efficacy of the carbon filtration process.

4. The flocculation process does not appear to be as important for the removal of pesticide as is the filtration process [based on the limited preliminary experimentation].

ACKNOWLEDGEMENTS

The authors wish to acknowledge the support of ICI Agrochemicals and ICI Americas Inc. for the initial funding of this research. Special thanks also go to Jim Mason for providing the GC analyses of the samples and Jane Cooper for the laboratory study on flocculation.

REFERENCES

[1] Ohio pesticide newsletter, A.C.Waldron, ed, August 28, 1990, pp 1-4.

[2] <u>Ground Water, Journal of the Association of Ground Water Scientists &
Engineers Volume 28, Number 2</u>, March-April 1990.

[3] EPA Executive Summary for proposed 40 CFR Part 165   J.K.Jensen, May
1990, pp 1-10.

[4] Instruction manual and spare parts list for the SENTINEL[*] effluent
treatment plant, E.Allman & Company Ltd, 17pp, 1989.

[5] Assessment of the Sentinel[*]/Carbo-flo[*] water effluent treatment,
Report No. C/87/0254, ADAS Research and Development Services 1987.

Baruch S. Shasha and Michael R. McGuire

STARCH MATRICES FOR SLOW RELEASE OF PESTICIDES

REFERENCE: Shasha, B. S. and McGuire, M. R., "Starch Matrices for Slow Release of Pesticides," Pesticide Formulations and Application Systems: 11th Volume, ASTM STP 1112, Loren E. Bode and David G. Chasin, Eds., American Society for Testing and Materials, Philadelphia, 1992.

ABSTRACT: Physical entrapment of both biological and chemical pesticides was acheived by encapsulation of active ingredients within starch matrices. Starch is first gelatinized either physically using heat, or chemically using alkali or urea, then pesticide is added followed by the crosslinking of starch. The final product is granular, ready to be used with conventional equipment. Electron microscope photographs reveal the presence of a honeycomb matrix containing the pesticide. Sprayable formulations based on entrapment within starch were also acheived. The sprayable material has a low and stable viscosity and can be used with conventional farm equipment. Upon application it dries to a thin film that does not dissolve in the presence of water.

KEYWORDS: starch, encapsulation, biological pesticides

Formulation research utilizing cornstarch at the Northern Regional Research Center has resulted in several techniques that allow for sustained release of pesticides [1-8]. Early work, to encapsulate chemical herbicides, involved the chemical modification of starch with strong urea or alkali solutions. Following modification, the starch was cross-linked in the presence of the active agent to yield water-insoluble particulates. Alternatively, the starch could be treated with high heat and then cross-linked to yield a similar product. These granules can be applied using existing farm equipment, are safer to use than conventional formulations, and have the potential to reduce chemical leaching into groundwater. The rate of release of

Drs. Shasha and McGuire are research chemist and research entomologist, respectively, in the Plant Polymer Research Unit, Northern Regional Research Center, Agricultural Research Service, U. S. Department of Agriculture, Peoria, IL 61604 (U.S.A).

active ingredients is dependent on factors such as physical characteristics of the pesticide, presence of moisture, and the presence of enzymes such as amylases that degrade the starch moiety. Preliminary field tests with herbicides encapsulated as above indicate sustained release and high activity while reducing the amount of active ingredient.

Recently, formulation of biological control agents has received considerable attention.   Current public concerns about pesticide usage has accelerated research of environmentally acceptable control measures.   One of the biggest hinderences to using these living agents is their short survival time in the field following application. Formulation will play a major role in the successful and widespread use of non-chemical means of pest control.

While the encapsulation processes outlined above for chemical pesticides are generally too harsh for biological control agents, starch still can be used for their formulation.   The rest of this paper will be concerned with the formulation of insect pathogens; however, weed and disease biological control agents may also be formulated using the techniques described below.

GRANULAR FORMULATIONS

Dunkle and Shasha [9] encapsulated <u>Bacillus</u> <u>thuringiensis</u> Berliner (Bt) in a starch matrix for use against the European corn borer.   The procedure involved mixing a pregelatinized starch such as Mira-gel (Staley Inc, Decatur, IL) with a mixture of water, Bt, and corn oil.   A gelatinous mass formed and, after several hours, became non-sticky. The mass was then chopped in a blender with a small amount of pearl starch to yield particles that passed 20 mesh.   The entire process was conducted at room temperature or colder and did not involve the use of harsh chemicals.   Upon ingestion, the starch granule is degraded by α-amylases present in the guts of phytophagous insects thus releasing the Bt.   To determine whether the encapsulation was causing inactivation of Bt, granules were treated with amylase to release the Bt and then incorporated into diet for assay against corn borer larvae.   No differences were observed in mortality between diet treated with digested starch-encapsulated Bt and diet treated with Bt that had not been encapsulated.

Like most microbial control agents, unformulated Bt will not survive well in the environment.   One factor in particular, ultraviolet (UV) light is especially damaging [10].   When suitable UV screens such as congo red or folic acid are added to the starch formulation Bt activity can be extended up to 12 days.   Starch-encapsulated Bt lacking the protectants lost all activity within four days exposure to direct sunlight (Figure 1).   Similarly, additives to enhance insect feeding on the granules can be incorporated.   These compounds, such as Coax (CCT Corporation, Litchfield Park, AZ) cause increased feeding on the granules even in the presence of Bt; thus insects ingest more toxicant.   Greenhouse [11] and field [12] tests demonstrated that relatively low amounts of Bt could be added to the starch formulation without loss of efficacy when Coax was also added (Figure 2).

The technique to encapsulate Bt within starch matrices may also be used to encapsulate grasshopper entomopoxviruses.   One of these

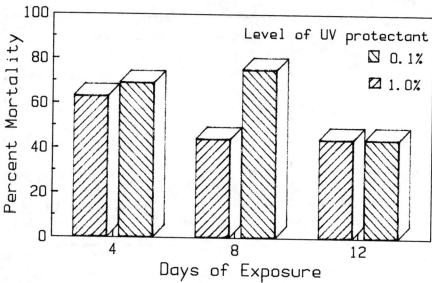

Figure 1. Starch granules containing Bt and congo red or folic acid were exposed to direct sunlight and then fed to neonate European corn borers. Results are averaged over the two additives. Bt in granules containing no UV protectant caused no mortality after four days. After [10].

TABLE 1 -- Effect of entomopoxvirus formulation on grasshopper percent infection[a]

Formulation Number	Source of oil	Molasses	UV Screen	Percent Infection
1.	Corn	-	-	61.0
2.	Corn	-	Congo Red	30.1
3.	Wheat Germ	-	Congo Red	35.6
4.	Wheat Germ	+	Congo Red	44.2
5.	Wheat Germ	+	Carbon	89.7
6.	Wheat Germ	-	-	64.5
7.	Wheat Germ	-	Congo Red	40.3
8.	Wheat Germ	+	-	61.7
9.	Wheat Germ	+	Congo Red	38.9
10.	Wheat Germ	+	Carbon	86.1
Wheat Bran Bait[b]				75.8
Control[c]	Wheat Germ	-	-	1.5

[a]Grasshoppers were exposed to granules for 3 d then held for 21 d. Infection determined by microscopic analysis.
[b]Virus sprayed onto flaky wheat bran (commercial standard).
[c]No virus in formulation.

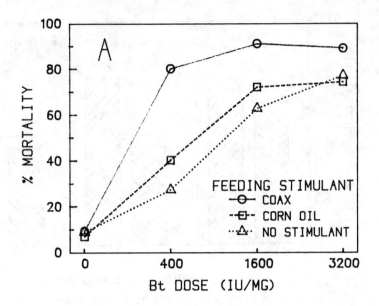

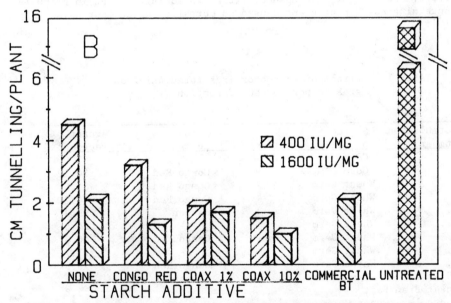

Figure 2. (a) Starch granules were formulated with one of three levels of Bt and one of three feeding stimulant treatments and placed in the whorls of corn plants in the greenhouse. European corn borer egg masses were pinned to the whorls and allowed to hatch. Four days later plants were dissected and percent mortality readings were taken. After [11].

(b). Starch granules containing one of two levels of Bt and one of four additive treatments were prepared and assayed in field corn for efficacy against European corn borer. A commercial formulation of Bt was also evaluated. Numbers represent average cm tunnelling over four replications. After [12].

viruses, isolated from <u>Melanoplus sanguinipes</u>, is currently under evaluation for grasshopper control. McGuire et al. [13] determined that the starch encapsulation technique can be successfully used to formulate this virus. A UV protectant (charcoal) and a feeding stimulant (molasses) are necessary for an efficaceous formulation. This formulation caused up to 90% mortality in laboratory bioassays but formulations lacking one or more of the ingredients caused significantly less mortality (Table 1). Starch provides an ideal vehicle for the incorporation of all necessary ingredients. In a field test, starch-encapsulated virus was applied to a 20 ha plot of rangeland infested with grasshoppers. Seven d later, up to 60% of the population was infected with virus, while grasshoppers in the control plot remained healthy.

To produce quantities of starch granules necessary for field tests such as the one conducted on grasshopper-infested rangeland (eg 100s of pounds), a pilot-scale process was developed. In the laboratory, five-pound batches are relatively easily made using an 8-1 double planetary mixer (Charles Ross and Son Co. Hauppauge, NY). This product is chilled and ground in a blender to acheive the desired particle size. We discovered that the extra addition of pearl starch during grinding is not necessary if the gelatinous mass is allowed to stand for at least one hour at 10°C. Initial attempts at making larger batches of granules involved the use of a "granulator" manufactured by APV Co. (Saginaw, MI). A 60 L kettle could produce approximately 9 kg of large-sized particles (ca 10mm diameter) in approximatel 20 min. However, the granulator was expensive and larger batch sizes were necessary. After further attempts with other equipment, we found that a 6 cu ft. mortar mixer could be used to mix 23 kg batches successfully. The final formulation consisted of water (23 1), molasses (4 1), oil (1 1), charcoal (100g), and the virus mixture. These ingredients were allowed to mix for several minutes and then 23 kg of Mira-gel was added. Following a 15 min period of mixing, the mixer was shut off and allowed to stand one hour at approximately 15°C. The mixer was then turned on and the rotating blades broke the mass into particles of 10-20 cm in diameter. These particles were then passed through a single-disc mill, dried, and stored at 4°C until use. Encapsulation efficiency with this method as measured by the amount of oil on the outside was approximately 85%. Approximately 230 kg of formulated entomopoxvirus was made in two days.

Recently, Carr et al. [14] reported a method for continuous production of starch-encapsulated herbicide. This process, utilizing inexpensive pearl starch in a twin-screw extruder at 65°C, was used to successfully encapsulate Bt. The process resulted in the Bt being exposed to this high temperature for only a few seconds. Bioassay [15] of the extruded product yielded 70% mortality in neonatate corn borers while the product made with pregelatinized starch at room temperature [9] yielded 71% mortality. The possibility of encapsulating other insect pathogens has not been explored.

SPRAYABLE FORMULATIONS

Although work with starch-encapsulated granular products has a long history, sprayable formulations capable of encapsulating active

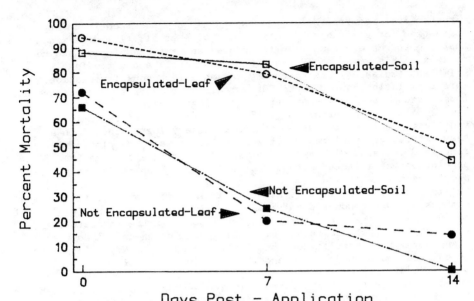

Figure 3.   A sprayable starch formulation of Bt was applied to cotton
leaves and either watered over the leaves or watered just to the soil
in the greenhouse. At specified time intervals, leaves were excised
and placed in a dish with neonate European corn borers. Three days
later, dead and live larvae were counted. Numbers represent average
percent mortality based on ten dishes/date/treatment.

ingredients have only recently been developed. Microencapsulated
biopesticides manufactured by Lim Technology Laboratories have the
advantage of being able to incorporate UV screens to extend activity.
Mycogen Corporation has developed the Cell Cap system that also can
extend insecticidal activity. While these technologies represent
unique methods for blocking UV, the products are still susceptible to
wash-off from rainfall. A formulation developed recently, however,
utilizes soluble pre-gelatinized starch that entraps active ingredients
on the leaf surface after application. This process results in a
product that can resist washoff and is versatile enough to allow for
incorporation of UV screens and feeding stimulants [16]. The
formulation is composed of pre-gelatinized starch and sucrose premixed
in equal amounts. The sucrose acts to physically separate and disperse
the starch particles upon mixing in water and it acts as a sticker on
the plant. Active ingredients such as Bt, UV screens, or feeding
stimulants can either be premixed with the dry ingredients or can be
added to the spray tank separately from the starch/sucrose mixture.
Once the formulation is applied, an insoluble film is formed thus
entrapping active ingredients on the leaf surface. The film is
permeable to gasses but resists washoff under greenhouse artificial
rain situations. Bioassays indicate that this formulation may also act
to reduce the effect of leaf factors that may degrade Bt. Bt applied
directly to cotton leaves loses activity within 7 days even when the
leaves were kept dry. Bt applied in the starch formulation, however,
retains high activity over a two week period (Figure 3).

SUMMARY

    Starch encapsulation is an extremely versatile system that may be adapted to meet the needs of many agricultural systems. Starch is widely abundant, inexpensive as an encapsulating agent, and easily modified. The granular products show good activity whether chemical or biological control agents are used. Extended activity and use of less active ingredients are but two of the many attributes of granular formulations. Sprayable formulations, though recently developed, have promise for much wider application systems. Many more agricultural products are applied with spray systems than with granule systems thus leading to a larger market potential.

REFERENCES

[1]   Shasha, B.S., Doane, W.M., and Russell, C.R. "Starch-Encapsulated
        Pesticides for Slow Release", Journal of Polymer Science,
        Polymer Letters Edition, Vol. 14, 1976, pp. 417-420.
[2]   Shasha, B.S., Doane, W.M., and Russell, C.R.,"Encapsulation by
        Entrapment", U.S. Patent 4,344,857, 1982.
[3]   Shasha, B.S., Trimnell, D. and Otey, F.H., "Encapsulation of
        Pesticides in a Starch-Calcium Adduct" Journal of Polymer
        Science, Polymer Chemical Edition, Vol. 19, 1981, pp
        1891-1899.
[4]   Trimnell, D., Shasha, B.S., Wing, R.E., and Otey, F. H.,
        "Pesticide Encapsulation Using a Starch-Borate Complex as Wall
        Material", Journal of Applied Polymer Science, Vol. 27, 1982,
        pp. 3919-3928.
[5]   Shasha, B.S., Trimnell, D. and Otey, F.H., "Starch-Borate
        Pesticides for Slow Release", Journal of Applied Polymer
        Science, Vol. 29, 1984, pp. 67-73.
[6]   Riley, R. T., "Starch-Xanthate-Encapsulated Pesticides: A
        Preliminary Toxicological Evaluation", Journal of Agricultural
        Food Chemistry, Vol. 31, 1983, pp. 202-206.
[7]   Trimnell, D. and Shasha, B.S. "Controlled Release Formulations of
        Atrazine in Starch for Potential Reduction of Groundwater
        Pollution", Journal of Controlled Release, Vol. 12, 1990, pp.
        251-256.
[8]   Trimnell, D., Shasha, B.S. and Otey, F.H., "The Effect of
        alpha-Amylases Upon the Release of Trifluralin Encapsulated in
        Starch" Journal of Controlled Release, Vol. 1, 1985, pp.
        183-190.
[9]   Dunkle, R. L. and Shasha, B. S., "Starch-Encapsulated Bacillus
        thuringiensis: A Potential New Method For Increasing
        Environmental Stability of Entomopathogens", Environmental
        Entomology Vol. 17, 1988, pp. 120-126.
[10] Dunkle, R. L. and Shasha, B. S., "Response of Starch-Encapsulated
        Bacillus thuringiensis containing Ultraviolet (UV) Screens to
        Sunlight", Environmental Entomology, Vol. 18, 1989, pp.
        1035-1041.

[11] Bartelt, R. J., McGuire, M. R., and Black, D. A.," Feeding
      Stimulants for the European Corn Borer
      (Lepidoptera: Pyralidae): Additives to a Starch-Based
      Formulation for Entomopathogens". Environmental Entomology,
      Vol. 19, 1989, pp. 182-189.
[12] McGuire, M. R., Shasha, B. S., Lewis, L. C., Bartelt, R. J., and
      Kinney, K., "Field Evaluation of Granular Starch Formulations
      of Bacillus thuringiensis Against Ostrinia nubilalis
      (Lepidoptera: Pyralidae)", Journal of Economic Entomology, in
      press.
[13] McGuire, M. R., Streett, D. A., and Shasha, B. S., "Evaluation of
      Starch-Encapsulation For Formulation of Grasshopper
      (Orthoptera: Acrididae) Entomopoxviruses", Journal of Economic
      Entomology, in preparation.
[14] Carr, M. E., Doane, W. M., Wing, R. E., and Bagley, E. B.,
      "Starch Encapsulation of Biologically Active Agents by a
      Continuous Process", U. S. Patent Application Serial Number
      07/542,566, 1990.
[15] Hughes, P. R. and Wood, H. A., "A Synchronous Peroral Technique
      for the Bioassay of Insect Viruses", Journal of Invertebrate
      Pathology, Vol. 37, 1981, pp. 154-159.
[16] McGuire, M. R. and Shasha, B. S., "Sprayable Self-Encapsulating
      Starch Formulations for Bacillus thuringiensis", Journal of
      Economic Entomology, in press.

Robert E. Wing, Merle E. Carr, William M. Doane, and Marvin M.
Schreiber

STARCH ENCAPSULATED HERBICIDE FORMULATIONS: SCALE-UP AND LABORATORY
EVALUATIONS

REFERENCE: Wing, R. E., Carr, M. E., Doane, W. M., and Schrieber,
M. M., "Starch Encapsulated Herbicide Formulations: Scale-Up
Laboratory Evaluations," Pesticide Formulations and Application
Systems: 11th Volume, ASTM STP 1112, Loren E. Bode and David G.
Chasin, Eds., American Society for Testing and Materials,
Philadelphia, 1992.

ABSTRACT: Non-chemically modified cornstarch effectively
encapsulates herbicides. Thousand pound quantities of
atrazine, alachlor and metolachlor were prepared by extrusion
followed by drying, grinding, and sieving to give the desired
size granules. Leaching potential of these formulations was
evalulated using soil columns and bioassay techniques. Large
quantities of products were sent to cooperators in 10 states
for evaluation along with commercial formulations.

KEYWORDS: starch, encapsulation, herbicides, slow release,
groundwater

Previously, it was shown that encapsulation of pesticides was
achieved by dispersing the pesticides into an aqueous dispersion of
gelatinized starch and then crosslinking the starch by xanthide [1];
calcium chloride [2]; borate [3]; or calcium-borate [4] methods.
These studies emphasized improving worker saftey in handling
pesticides and providing a release profile for soil incorporated
herbicides such that the release rate of active ingredient was
sufficient to control the target weed but slow enough to avoid
phytotoxicity to the crop.

Several cooperators have shown that starch encapsulation: a)
significantly decreases percutaneous permeability of parathion as

Drs. Wing, Carr, and Doane are research chemists at Northern
Regional Research Center, Agricultural Research Service, U. S.
Department of Agriculture, Peoria, IL 61604 (U.S.A.) and Dr. Schreiber
is a research agronomist at Agricultural Research Service, U. S.
Department of Agriculture, Purdue University, West Lafayette, IN 47907.

compared with clay formulated samples [5]; b) extends the duration of weed control by reducing release rates [6-7]; c) obviates the need for soil incorporation of certain herbicides [8]; and d) reduces amounts of herbicide required [9].

Starch exhibits many properties required of a polymer to function as an encapsulation matrix. Broader acceptance of starch as an encapsulation matrix depends on improving the economics of encapsulation and/or eliminating the use of chemicals to form the cross-linked matrix. Recent articles [10-13] and a patent [14] showed that herbicides are effectively encapsulated in starch without the use of crosslinking agents. Cornstarch is first solubilized or highly dispersed by steam injection cooking and then crosslinked through the natural process of retrogradation after the herbicide is added. Retrogradation is important in gelatinized starch and defines the formation of aggregates resulting from hydrogen bonding between hydroxyl groups of adjacent starch chains. Factors such as rate of cooling, pH, amylose content, moisture content, and dispersion temperature affect retrogradation of solubilized starch. This property is beneficial for enhancing the encapsulation and controlling the release rate of pesticides from the starch matrix. Mixtures of various starches (waxy, pearl, and high amylose) can be selected to control the amylose content and thus the degree of retrogradation. Several cooperators [15-16] have reported excellent results in controlling weeds and reducing groundwater contamination with products prepared by this encapsulation technique [17].

Recently we produced starch encapsulated products in a continuous, efficient, and effective process by using a twin-screw extruder to gelatinize starch and incorporate the pesticide [18-19]. A preblend of starch and atrazine was fed into the twin-screw extruder while water was injected to gelatinize starch allowing the atrazine to enter the matrix. Other herbicides (alachlor or metolachlor) were metered into other formulations as liquids. The extrudates were dried and ground to provide products with 94-99% atrazine, 72-95% alachlor, and 87-92% metolachlor encapsulated. We now report a scale-up of the extrusion process to produce thousand-pound quantities of these starch encapsulated herbicides. Preliminary data of their evaluation in greenhouse bioassay and soil columns will be presented.

EXPERIMENTAL METHODS

a. Chemicals

Starch from CPC International, Englewood Cliffs, NJ was used. Atrazine [2-chloro-4-ethylamino-6-isopropylamino-s-triazine (97%)], technical grade, metolachlor [2-chloro-2'-ethyl-6'-methyl-N-(1-methyl-2-methoxyethyl)acetanilide (95%)] technical grade and Dual 8E (86% metolachlor) were supplied by Ciba Geigy Corp., Greensboro, NC. Alachlor [2-chloro-2',6'-diethyl-N-(methoxymethyl)acetanilide (94.3%) technical grade was supplied by Monsanto Co., St. Louis, MO. Alachlor was melted at 50°C before use.

b. Extrusion and encapsulation

The extruder used was a ZSK 57 corotating, fully intermeshing twin-screw extruder (Werner and Pfleiderer). The barrel length/screw diameter (L/D) ratio was 30. Starch was fed into barrel section (BS) 1 at the rate of 150 lb/hr (135 lb/hr dry basis). Atrazine as a solid was also fed into BS-1 at the 10% addition level based on starch, while in other formulations Dual or melted alachlor were metered into BS-3. Barrel temperatures were 25°C at BS-1, 75°C at BS-2 to BS-4, and 95°C at BS-5 to BS-10. The screw at 200 rpm was composed of alternating conveying and kneading block elements in the starch gelatinization zones. The die head assembly was equipped with a die having twenty 5-mm diameter holes. Production rate was 100 kg/hr product with 30% moisture. Products were cut with a die-face cutter at 70% solids at the extruder exit. The extrudate was dried and ground to the desired particle sizes in a Bauer disc mill. The milled samples (2 to 25%) moisture were sieved to obtain 14-20 or 20-40 mesh products.

c. Percent herbicide encapsulated

The amount of herbicide entrapped in the sieved products was determined for samples (100 g) washed three times with chloroform (300 ml total) to remove absorbed herbicide. Products were analyzed for % active agent (a.i.) via nitrogen analysis for atrazine and chlorine analysis for metolachlor and alachlor as previously discussed [18-19].

d. Swellability

Samples (0.20 g, 20-40 mesh) were placed in a 10-ml graduated cylinder with water (4.0 ml) at 30°C and gently stirred several times during the first three hours to prevent clumping. After 24 h, the height of the swollen product in the cylinder was used to calculate the % increase in volume.

e. Soil column leaching studies

Dry-screened (1/2 cm hardware cloth) Miami silt loam top soil was packed into 7.5 cm dia. aluminum tubes. Each tube had a 2.5-cm wide slot, 40-cm long down one side. This slot was covered and sealed with floral clay to produce a water-tight seal. The dry soil was scooped up and poured into the column. After each scoop was added the entire column was dropped 4 times onto a 7.5-cm dia. rubber stopper from a height of 5 cm. This procedure was repeated until the soil reached 2 cm above the top of the slot. The columns were saturated with ~750 ml water. Starch-encapsulated products (20-40 mesh) and commercial EC samples were applied to the top of the columns at the rate of 3.36 kg/ha. Sand (1 cm) was placed on the samples and 375 ml water (equivalent to 7.5 cm rainfall) was passed through the column at a rate of 4 ml/min. A 1-cm head of water was always maintained during the leaching period to prevent channeling. After leaching, the column was

placed on its side and the slot was opened. Foxtail seeds (~200) for
alachlor and metolachlor and bentgrass for atrazine were sown on the
exposed soil and a 0.5-cm layer of sand was added to cover the
seeds. The sand was kept slightly moist during the 14-day growing
period. The distance from the top of the slot to where the growth
started determined the depth of leaching.

f. Herbicide release studies into water

    Samples (20-40 mesh) of encapsulated atrazine, alachlor, or Dual
(100 mg) in water (70 ml) were placed in 125-ml Erlenmeyer flasks.
Mixtures were agitated in an orbital shaker (Lab-Line, Melrose Park,
IL) at 100 shakes per minute and sampled at 1, 2, 3, and 21 h. Water
volume was in large excess of that required to dissolve all of the
herbicides (atrazine solubility is 33 mg/ml water at $27^{o}C$, alachlor
solubility is 240 mg/ml at $23^{o}C$, and metolachlor solubility is 530
mg/ml at $20^{o}C$-Herbicide Handbook. Herbicide concentrations in the
water phase were determined spectrophotometrically (Beckman DU-50,
Irvine, CA) at their ultraviolet absorption maxima. Corrections were
made at these wavelengths using starch controls containing no
herbicide.

RESULTS

    Large quantities (750 kg) of starch encapsulated atrazine,
alachlor, and metolachlor were prepared on a ZSK 57 extruder. Table
1 shows that these herbicides were encapsulated effectively in the
scale-up at a production rate of 70 kg/h. The swellabilities listed
in Table 1 are an artificial indication of how fast the herbicides
will be released. Portions of the products were ground at different
moisture contents after either air or oven drying. Drying the samples
prior to grinding improved the encapsulation process. Therefore it is
important during continuous, commercial scale processing that the
products are dried to about 10% moisture prior to grinding. In Table
2 the release data in a laboratory assay are reported.

    The starch encapsulated herbicides were compared with the
commercial EC formulations in soil column tests using 7.5 cm simulated
rainfall. The results in Table 3 show a significantly reduced movement of active agent from encapsulated products. All the encapsulated
samples were unwashed so the surface herbicide was still present.

    Field testing of the new formulations was initiated in 10 states
in the midwest during the 1990 growing season and is still in progress
at this time.

CONCLUSIONS

    Using a twin-screw extruder for encapsulating herbicides in
starch matricies is efficient, effective and continuous. Laboratory

evaluation of the products shows them to be slow release, effective in controlling weeds, and able to reduce leaching under controlled conditions.

TABLE 1 -- Scale-up encapsulation of atrazine, alachlor, and metolachlor[a]

Active agent	% Water at grinding	Drying method	14-20 Mesh		20-40 Mesh		% swell-ability in water[c]
			% total active agent	% encapsu-lation[b]	%total active agent	% encapsu-lation[b]	
atrazine	4	air	11.3	94	11.1	95	340
alachlor	16.5	air	9.6	80	9.6	66	220
alachlor	7.8	air	9.4	93	9.5	91	180
alachlor	2.7	oven[d]	10.0	90	9.7	90	180
alachlor	2.3	oven	9.7	95	9.8	89	220
metolachlor	25	air	9.2	50	9.4	34	200
metolachlor	18	air	9.2	41	9.2	29	300
metolachlor	10	oven	9.0	70	9.0	55	180
metolachlor	7	air	8.7	77	8.9	62	180

[a]Starch at 70% concentration, ~ 10% herbicide.
[b]Herbicide remaining after extracting surface material.
[c]Sample (0.2 g) in water (4 ml).
[d] Dried at 50°C.

TABLE 2 -- Rate of release of encapsulated herbicides into water[a]

Product encapsulated	% Released			
	1 hr	2 hr	3 hr	21 hr
atrazine	4	7	12	40
alachlor	11	18	23	46
metolachlor	43	58	68	100

[a]20-40 mesh unwashed samples (100 mg) in 70 ml of water swirled at 100 rpm.

TABLE 3 -- Soil column leaching of herbicides from starch and commercial EC formulations[a]

Herbicide formulation	Depth, cm
atrazine	
starch	6
EC	33
alachlor	
starch	5
EC	24
metolachlor	
starch	5
EC	27

[a]Starch (20-40 mesh) and EC formulations at 3.36 kg/ha and 7.5 cm rainfall.

REFERENCES

[1] Wing, R. E. and Otey, F. H.,"Determination of Reaction Variables for the Starch-Xanthide Encapsulation of Pesticides", Journal of Polymer Science, Polymer Letters Edition, Vol. 21, 1983, pp.121-140.

[2] Shasha, B. S., Trimnell, D., and Otey, F. H., "Encapsulation of Pesticides in a Starch-Calcium Adduct", Journal of Polymer Science, Polymer Chemical Edition, Vol. 19, 1981, pp. 1891-1899.

[3] Trimnell, D., Shasha, B. S., Wing, R. E., and Otey, F. H.,"Pesticide Encapsulation Using a Starch-Borate Complex as Wall Material", Journal of Applied Polymer Science, Vol. 27, 1982, pp. 3919-3928.

[4] Wing, R. E., Maiti, S., and Doane, W. M., "Factors Affecting Release of Butylate from Calcium Ion-Modified Starch-Borate" Journal of Controlled Release, Vol 5, 1987, pp. 79-89.

[5] Reily, R. T., "Starch-Xanthate-Encapsulated Pesticides: A Preliminary Toxicological Evaluation, Journal Agricultural Food Chemistry, Vol. 31, 1983, pp. 202-206.

[6] Schreiber, M. M., Shasha, B.S., Ross, M. A., Orwick, P. L., and Edgecomb, D. W., Jr., "Efficacy and Rate of Release of EPTC and Butylate from Starch Encapsulated Formulations Under Greenhouse Conditions", Weed Science, Vol. 26, 1978, pp. 679-686.

[7] Devisetty, B. N., McCormick, C. L., and Shasha, B. S., "Controlled Release Herbicides", Proceedings of the 7th International Symposium on Controlled Release of Bioactive Materials, Fort Lauderdale, FL., 1980, pp. 187-188.

[8] White, M. D. and Schreiber, M. M., "Herbicidal Activity of Starch Encapsulated Trifluralin", Weed Science, Vol. 32, 1984, pp. 387-392.

[9] Coffman, C. B. and Gentner, W. A., "Herbicidal Activity of Controlled Release Formulations of Trifluralin", Indian Journal of Agricultural Science, Vol. 54, 1984, pp. 117-122.

[10] Wing, R. E., Maiti, S., and Doane, W. M., "Effectiveness of Jet-Cooked Pearl Cornstarch as a Controlled Release Matrix", Starch, Vol. 39, 1987, pp. 422-425.

[11] Wing, R. E., Maiti, S., and Doane, W. M., "Amylose Content of Starch Controls the Release of Bioactive Agents", Journal of Controlled Release, Vol. 7, 1988, pp. 33-37.

[12] Wing, R. E., "Non-chemically Modified Cornstarch Serves As An Entrapment Agent", Proceedings of the Corn Utilization II, Vol. 2, 1988, pp.1-17.

[13] Wing, R. E., "Cornstarch Encapsulated Herbicides Show Potential to Reduce Groundwater Contamination", Proceedings of the 16th International Symposium on Controlled Release of Bioactive Materials, Chicago, IL, 1989, pp. 430-431.

[14] Doane, W. M., Maiti, S., and Wing, R. E., "Encapsulation by Entrapment within Matrix of Unmodified Starch", U. S. Patent 4,911,952, 1990.

[15] Schreiber, M. M., Wing, R. E., Shasha, B. S., and White, M. D., "Bioactivity of Controlled Release Formulations of Herbicides in Starch Encapsulated Granules", Proceedings of the International 16th Symposium on Controlled Release of Bioactive Materials, Basel, Switzerland, Vol. 15, 1988, pp. 237-242.

[16] Schreiber, M. M., White, M. D., Wing, R. E., Trimnell, D., and Shasha, B. S., "Bioactivity of Controlled Release Formulations of Starch-Encapsulated EPTC", Journal of Contolled Release, Vol. 7, 1988, pp.237-242.

[17] Schoppet, M. J., Gish, T. J., and Helling, C. S., "Effect of Starch Encapsulation on Atrazine Mobility in Small Undisturbed Soil Columns", Proceedings of the American Society of Agricultural Engineers, New Orleans, LA, 1989, pp. 1-13.

[18] Carr, M. E., Wing, R. E., and Doane, W. M., "Encapsulation of Atrazine Within a Starch Matrix by Extrusion Processing", Cereal Chemistry, (submitted).

[19] Wing, R. E., Carr, M. E., Trimnell, D., and Doane, W. M., Comparison of Steam Injection Cooking Versus Twin Screw Extrusion of Pearl Cornstarch for Encapsulation of Chloroacetanilide Herbicides", Journal of Controlled Release, (submitted).

Jenny A. Stein, William B. Kallay, G. R. Goss, and
Loukia K. Papadopoulos

A METHOD TO MONITOR RELEASE OF AN INSECTICIDE FROM
GRANULES INTO SOIL

REFERENCE: Stein, J. A., Kallay, W. B., Goss,
G. R., and Papadopoulos, L. K., "A Method to
Monitor Release of an Insecticide From Granules
into Soil," Pesticide Formulations and Application
Systems: 11th Volume, ASTM STP 1112, Loren E. Bode
and David G. Chasin, Eds., American Society for
Testing and Materials, Philadelphia, 1992.

ABSTRACT: A method has been developed to monitor
the release of an insecticide from granules into
soil. Pesticide release from granules and
subsequent movement into the soil is an important
parameter in pesticide performance. To date, no
quantitative measurements have been made. Past
studies have concentrated on either bio-assays or
content in the soil solution. Many field efficacy
studies have been performed. The described method
measures quantitatively release of an insecticide
into the soil environment in the laboratory. This
method consists of placing pesticide containing
granules into a pulverized soil. This mixture is
then incubated. Periodically the samples are
removed, the clay granules separated from the soil
and the soil analyzed for pesticide content. Using
this technique the effects of temperature, time,
soil moisture, carrier type, etc. are able to be
monitored for their effects on pesticide release.
The method and some of these effects are discussed.

KEYWORDS: pesticide, controlled release, granules

As a result of our commitment to the agricultural
industry, Oil-Dri has participated in the ASTM. In the
past, information has been presented on the theory, use
and methods involving granular carriers. Moll [1]
discussed the surface acidity of clays. Moll and Goss
[2] discussed the characteristics and uses of granular
agricultural chemical carriers. In 1989, Goss and
Reisch [3] presented a method to measure dust. This
paper discusses a method to monitor pesticide release
from granules into the soil.

Scientist, Scientist, Director, Product Technology,
Technician, respectively, Oil-Dri Corporation of
America, Chicago, IL 60611.

The subject of pesticide release is of paramount importance in formulations. Without the proper release characteristics, effective pest control may not be realized, or an excessive amount of material may be used. "Controlled" release of pesticides has been an active area of research. An examination of the patent literature reveals 70 patents during the 8 year period from 1981-1988. During the 2 years, 1989 and 1990, another 70 patents were issued. These include both medical and agrochemical usage. The scientific literature has also greatly expanded in recent years. During the 1990 Symposium on Controlled Release of Bioactive Materials (Reno, NV, USA) over 250 papers were presented.

Soil pesticide concentrations have been almost exclusively monitored by bioassay or leaching studies. Wing, [4], provides a good example of this style of work. Release into aqueous solutions is a convenient method to monitor pesticide release from granules but is somewhat unrealistic. Jamil et al [5] has followed release from granules into soil by isotopic techniques and Philips, et.al. [6] have followed pesticide release from plugs placed in the soil. By using the following procedure the gross movement of an insecticide into the soil has been followed chemically from treated granules. Using this technique, soil pesticide content has been measured and preliminary data on the effects of temperature, time, moisture and particle size have been collected.

MATERIALS AND METHODS

Soil: Soil used was an Odell silt loam; a fine-loamy, mixed mesic Aquic Argiudoll. Its characteristics are listed in Table 1 and are fairly typical of the soils from northeastern Illinois, USA.

TABLE 1 -- Soil characteristics

Type: Odell silt loam
Classification: Fine - loamy, mixed, mesic Aquic
                 Argiudoll.

% Sand	26.2	
% Silt	66.3	
% Clay	7.5	
pH	7.2	
% Organic Matter	4.6	
C.E.C. (meq/100g)	45.8	
Pounds per acre	P	116
	K	388
	Ca	16450
	Mg	940
	NH_3	430

Water Capacity, %   61.8

Field Capacity, %   20.4

For experimentation, the soil was allowed to air-dry. It was then hammermilled and the <60 mesh (250 μ) fraction collected.

Moisture was adjusted to the desired experimental range by spraying on water, mixing and equilibrating in a sealed container for at least 24 hours.

Carrier: Carriers used were either 18/40 (-1 mm, + 0.425 mm) LVM-MS or 8/16 (-2.36 mm, + 1.18 mm) LVM-MS clay from Oil-Dri Corporation of America, Chicago, Illinois. These are montmorrillonite clays in the low volatile material (LVM) state from Northeastern Mississippi. One synthetic carrier composed largely of organic matter was compared to the 18/40 particle size.

Pesticide: The pesticide used was an insecticide of the phosphate ester classification.

Granular Formulation: Pesticide was sprayed evenly onto the clay surface in a rotating, inclined container (wide-mouthed bottle).   Target concentration was 15% a.i. (active ingredient). All formulations were assayed before use. No deactivator was used. After all the insecticide was applied, the jar was sealed and agitated for 45 minutes.

Soil/Pesticide Preparation: Once the soil was adjusted to the desired moisture, the prepared granular formulation was added. Soil and pesticide were agitated to provide even distribution. This was 0.9 g formulation per 100 g soil. This provided a 1,350 ppm theoretical insecticide concentration. Treated soil samples were then sealed in wide-mouthed glass bottles and stored at the desired temperature.

Sample Analytical Preparation: For analysis, two 10 g samples were withdrawn from the storage container. Clay was separated from soil by sieving over first a 40 mesh (425 μ) followed by a 50 mesh (300μ) screen.   The 40/50 fraction was discarded and represented less than 1% of the sample. Both the +40 mesh sample (clay) and -60 mesh sample (soil) were analyzed for pesticide content. The entire clay fraction and 5 g of the soil fraction were added to two separate 25 ml wide-mouth glass bottles for analysis. Into these bottles was added 20 ml dichloromethane containing 0.24% dibutylphthalate internal standard. The bottle was sealed and agitated on an orbital shaker for 20 minutes at 400 rpm. The bottle was then removed and centrifuged 4 minutes at 1,000 rpm. Supernatent was directly injected into a gas chromatograph. Data reported is an average value of two samples from one storage container.

Gas Chromatographic Analysis: A Perkin-Elmer Sigma 300 gas chromatograph was used equipped with a Supelco SPB 5 (non-polar), 30 m, 0.32 ID (inside diameter) capillary column and a flame ionization detector. Conditions were oven temperature = 230 C, detector and injector temperatures = 250 C.   Helium pressure was 12 p.s.i. (pounds per square inch, 8.4×10^3 kg/m^2), O_2 and H_2 pressure 20 p.s.i. (1.4×10^4 kg/m^2), and column flow rate 2.3 ml/min. A 1 μl aliquot of the centrifuged sample supernatant was injected.

Results and Discussion

In general, the method has proved useful in providing guidelines as to the effects of temperature, moisture, particle size and carrier type. Pesticide longevity could also be followed and corresponds well to published literature. The method is also a realistic laboratory technique as it can directly follow the concentration of a pesticide into the soil, not just an aqueous solution.

Certain limitations are also inherent in the procedure. Physically, they involve soil moisture and particle size. As a laboratory technique it is also a static system, not dynamic as a field study is. Chemically, while the method is able to measure pesticide movement from a granule to the soil environment, it cannot predict pesticide activity if it is necessary to be in the soil solution for optimum efficacy. If the pesticide were systemic, it may not predict root uptake either. One must suppose, however, that the likelihood of being active in solution or taken up by roots is much greater if the pesticide is in the soil environment.

Method reproducibility is good as is shown in Figure 1. This is a study where the same soil treatment is followed three separate times. Here the soil moisture is 7.4% and at a temperature of 77°F (25°C). The graph shows a typical curve. Pesticide is released into the soil from the granules and soil concentrations increase to a maximum. After that point, disappearance of the pesticide occurs more rapidly than continued release and soil pesticide concentration drops. At the time of target pest pressure, there is adequate pesticide and it thereafter decomposes.

The effects of soil moisture are shown in Figure 2. Placement of the pesticide in an oven-dried soil at 77°F (25°C) shows quick release with subsequent disappearance. With 7.5% water a quick release is also noted. Soil pesticide content reaches a higher level than at 0% water but disappears as quickly. With more moisture (11.8%) release is slowed, as is subsequent disappearance. Published literature agrees with this phenomenon. This may be one of the reasons comparatively poor control is observed during a dry year.

At moisture levels higher than 12% it became too difficult to physically separate the clay particles from the soil particles. This is one of the inherent limitations of the method. The soil becomes too "clumpy" for good separation by sieving. Limited experiments have shown promise by wet sieving

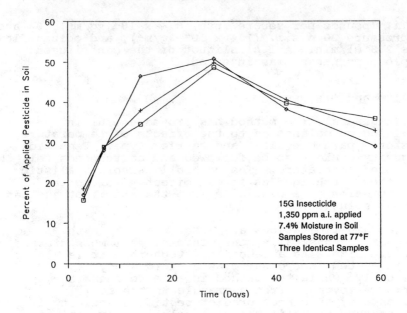

FIG. 1 -- Method Reproducibility

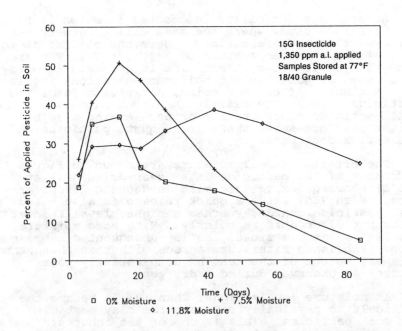

FIG. 2 -- Effect of Soil Moisture

techniques. For our initial studies we felt direct sieve separation is superior due to the unknown effects of large quantities of water on either the stability of the pesticide or leaching from the granules. The field capacity of 20.4% listed in Table 1 would correspond to a wilting point of about 8% water. Using this soil and an 18/40 particle size then restricts one to a somewhat limited range of applicable soil moistures.

The effects of temperature are listed in Figure 3. A soil temperature of 120°F (48.9°C) resulted in virtually instant release with subsequent disappearance of the pesticide. At 80°F (26.7°C) release and disappearance is tempered and consistent with pest pressure timing. A low temperature such as 40°F (4.4°C) results in a slow steady increase in soil concentrations throughout the 70 day period monitored.

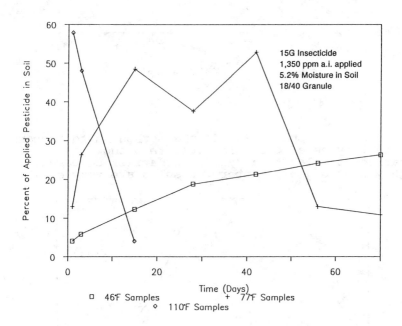

15G Insecticide
1,350 ppm a.i. applied
5.2% Moisture in Soil
18/40 Granule

□ 46°F Samples   + 77°F Samples
◊ 110°F Samples

FIG. 3 -- Effect of Temperature

Particle size effects are generally as expected. A comparison at 8% H$_2$O is shown in Figure 4. The 18/40 shows a greater initial release as expected and then disappears more quickly than the 8/16. Soil levels are more persistent, however, with 8/16. This same general effect was noted also at 0% and 12.8% moisture.

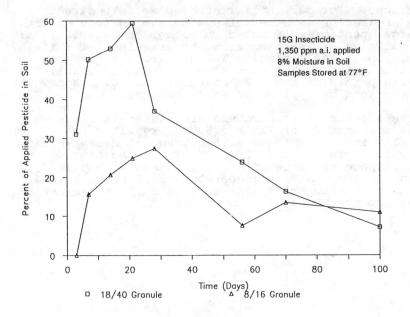

FIG. 4 -- Effect of Particle Size

One useful application of the method is to compare different types of carriers. Figure 5 shows a comparison of clay to a synthetic carrier composed of roughly half organic matter. Results are very similar.

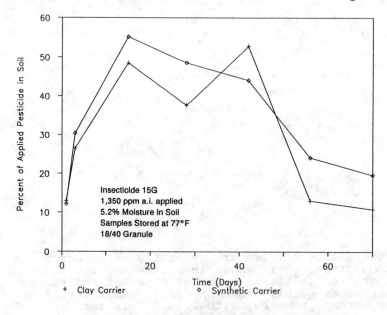

FIG. 5 -- Effect of Carrier Type

A materials reconciliation is shown in Figure 6.
The amount of pesticide found remaining in the separated
clay is added to the amount recovered from the soil and
compared to the originally applied amount of pesticide.
After an initial nearly complete reconciliation of
applied pesticide, (i.e. all of the pesticide was
accounted for) a decrease in pesticide recoverability
occurs. This is presumably due to degradation. We did
not follow degradation products, however. Little or no
pesticide was lost on the granule formulation itself.
This was followed as a control.

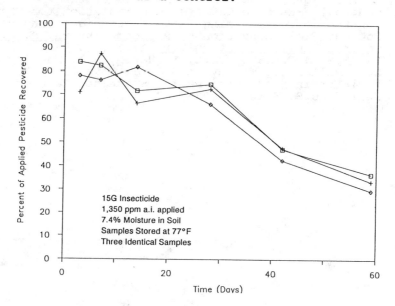

FIG 6. -- Materials Reconciliation

CONCLUSION

A method to monitor the release of an insecticide
from granules was developed. Useful applications
include the effects of moisture, temperature, particle
size and carrier type. Limitations are high soil
moisture levels and particle size separation.

REFERENCES

[1]   Moll, W. F., 1983, "Acidity Measurements for
      Carriers", Fourth Symposium on Pesticide
      Applications and Formulation Systems, ASTM, New
      Orleans, LA, Abstr.
[2]   Moll, W. F. and G. R. Goss, 1987, "Mineral Carriers
      for Pesticides - Their Characteristics and Uses",
      Pesticide Formulation/Application Systems, 6th

[3]   Goss, G. R. and F. J. Reisch, 1989, "A Technique for Dust Measurement", _Pesticide Formulation/Application Systems, 8th Vol.,_ Hovde/Beestman eds., ASTM (American Society for Testing and Materials) STP 980.

[4]   Wing, R.E., 1989, "Cornstarch Encapsulated Herbicides Show Potential to Reduce Groundwater Contamination", _Proceed. Intern. Symp. Control. Rel. Bioact. Mater. 16:430._

[5]   Jamil, F.F., M.J. Qureshi, A. Haq, N. Baskin and S.H.M. Naqvi, 1990, "Effectiveness of the Controlled-Release Carbofuran Formulation for Pest Control in Cotton using Isotope Techniques under Pot and Field Conditions", _Proceed. Intern. Symp. Control. Rel., Bioact. Mater._ 17:59.

[6]   Philips, J.C., H.L. Hoyt, L. Hazen, C.A. Macri, W.L. Miller, L. Wesley, and J. J. O'Neill, 1990, "Gel-Forming Compositions for Sustained Release in Soils", U.S. Patent 4,935,447.

Deborah H. Carter, Paul A. Meyers, and C. Lawrence Greene

TEMPERATURE - ACTIVATED RELEASE OF TRIFLURALIN AND DIAZINON

REFERENCE: Carter, D. H., Meyers, P. A., and Greene, C. L., "Temperature-Activated Release of Trifluralin and Diazinon," Pesticide Formulations and Application Systems: 11th Volume, ASTM STP 1112, Loren E. Bode and David G. Chasin, Eds., American Society for Testing and Materials, Philadelphia, 1992.

ABSTRACT: Landec Labs has been developing a family of polymers (Intelimer™) that when used in encapsulation, respond to a temperature change by releasing active ingredient (AI). Microcapsules made from these polymers containing selected pesticides have been developed which utilize this "off" and turn "on" property. Below the temperature of interest, the microcapsules have low release of the AI thereby protecting it from leaching and degradation. At the temperature of interest, the microcapsules have a higher release of the active ingredient. The change in release rate in response to temperature is unique to microcapsules made from Intelimer polymers and is not observed with commercial microcapsules. A variety of active ingredients including trifluralin and diazinon have been successfully microencapsulated using Intelimer polymers. Greenhouse and field studies indicate the microencapsulated trifluralin had reduced phytotoxicity to corn. In a cotton field trial, trifluralin microcapsules at a 0.55 kg AI/ha rate provided equivalent 90 day control of weeds to a 1.1 kg AI/ha treatment of Treflan 4EC™. Greenhouse results have demonstrated the temperature-activation characteristics with trifluralin and diazinon. The efficacy of diazinon capsules (made with Intelimer polymers) on corn root worm also will be discussed.

KEYWORDS: controlled release, temperature-activated, pesticides, diazinon, trifluralin

D. H. Carter is a project leader, P. A. Meyers is a formulation chemist and C.L.Greene is Director, Agricultural Products of Landec Labs, Inc., 3603 Haven Avenue,Menlo Park, California, 94025.

INTRODUCTION

Interest in controlled release technology in the agricultural arena continues to grow. Problems associated with groundwater contamination, degradation, volatilization and the banning of pesticides have contributed to the need for more efficient delivery systems [1]. Positive economic and environmental impact can be realized by more efficient delivery of active ingredients [2]. Predictable increases in soil temperature during the growing season triggers a variety of biological events such as germination, egg hatching, and pupation [3,4]. The most effective control of the pest population would be obtained by delivering the pesticide at the temperature at which the pest becomes active. If a commercially practical method could be developed to automatically deliver pesticides to crops when the target pests are emerging, greater control could be achieved with less active ingredient and fewer applications. The results would include reduced application rates, reduced crop phytotoxicity, extended efficacy and greater overall cost effectiveness. Negative environmental impact and ground water contamination could also be reduced. Landec Labs has developed a family of side chain crystallizable polymers with melting points in the 15°C-35°C range, which are temperature-activated and when used in microcapsule and granule formulations can protect agrichemicals until they are required [5].

Hence, it was the objective of this research to determine the potential benefits of Intelimer formulated pesticides under greenhouse and field conditions. Testing was conducted with the following objectives:

A. Reduce the amount of active ingredient required to achieve economic pest control.
B. Assess the ability of the Intelimer to reduce crop phytotoxicity.
C. Demonstrate temperature cycling of Intelimer microcapsules in water-based formulations.

MATERIALS AND METHODS

The polymers used in this work were prepared via solution polymerization. Polymers were purified by precipitation into a non-solvent and characterized by gel permeation chromatography, infrared spectroscopy, and thermal analysis. Side chain crystallizable polymers used in encapsulation have an acrylic backbone, with a series of long chain fatty alcohols esterified to it. These fatty alcohol moieties are referred to as "side chains". It is the ability of these side chains to crystallize and then melt over a very narrow temperature range (5 degrees Centigrade), which produce various release rates of the microcapsule as a function of temperature. See Figure 1 for a schematic representation of the process. The side chain length influences the release temperature of the polymer by changing the polymer melt point(T_M). Side chain lengths of 12 to 18 carbon units are used to achieve release temperatures from 15°C to 35°C.

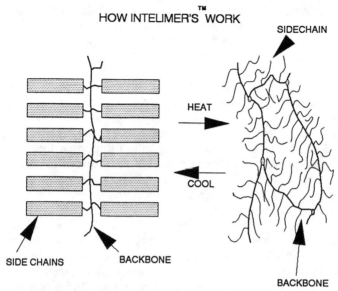

HOW INTELIMER'S™ WORK

SIDECHAIN

HEAT

COOL

SIDE CHAINS   BACKBONE

BACKBONE

• SIDE CHAINS CRYSTALLIZE TO FORM AN
  EFFECTIVE DIFFUSION BARRIER.

• WHEN WARMED, SIDE CHAINS MELT AND PLASTICIZE POLYMER,
  ALLOWING CHEMICALS TO DISSOLVE AND MIGRATE.

FIG. 1 -- Schematic representation of the process.

In order to demonstrate the versatility and potential effectiveness of the Intelimer polymers, a liquid insecticide, diazinon, and a low melting solid herbicide, trifluralin, were identified. Diazinon was chosen for the initial study since it is widely used in insect control in home and commercial agricultural applications. On the other hand, trifluralin controls annual grass and broadleaf weeds. Both active ingredients are well characterized in the literature. Technical grades of diazinon and trifluralin were supplied by Ciba Geigy and Griffin Corporation, respectively.

Microcapsules were prepared using standard emulsion processing. The active ingredient is incorporated into the oil phase, emulsified in water and crosslinked. The microcapsules consist of approximately 10% polymer based on the weight of active ingredient. Microcapsules were verified to have a diameter of 30 microns by optical and scanning electron microscopy and particle size analysis. For evaluation, the microcapsules were supplied in an aqueous formulation with no adjuvants, emulsifiers or other additives incorporated. Landec 15G granules were prepared by polymerizing the active ingredient (15 wt%) in the polymer and grinding the particles to a commercial size. Diazinon release rates were conducted in water. The concentration of available AI in the solution was determined using a UV spectrophotometer set at 246 nm. Trifluralin release rates were conducted in ethanol:water (1:1 V/V). The concentration of the AI in solution was determined using a UV spectrophotometer set at 279 nm. The water and the water:ethanol release mediums were simply used as a laboratory means to address how the release may be occurring and not to provide a correlation to soil release data or biological systems. Different release mediums were required for the different actives because of the poor water solubility of trifluralin in pure water. Release rates were varied by manipulating the polymer characteristics, particle size and crosslinking.

The northern corn rootworm (Diabrotica barberi) bioassay was conducted by Midwest Research Inc., in York, Nebraska. In 1989, the measured soil temperature at a 2" depth for this test site in June and early July ranged from 15°C to 35°C. Soil from the farm (with the following consistency: sand, 18.7%; silt, 57.3%; clay, 24%; pH, 6.3; CEC, 14.5, dry soil) was thoroughly mixed with each formulation and placed in cups. One partially germinated corn seed and 10 corn rootworm larvae were introduced to each cup which was capped. These cups were incubated at the appropriate temperature. After 48 hours, all the soil was sieved and the number of living corn rootworms was determined.

The trifluralin greenhouse and field bioassays in corn were conducted by Research for Hire, Inc. in Porterville, California. For the greenhouse bioassay, the formulations were sprayed on the soil, thoroughly mixed with it, then put into bread pans. The soil type was sandy loam with a pH of 7.3 and 0.7% organic matter. Application rates of 0.21, 0.42, 0.83 and 1.6 kilograms of active ingredient per hectare were used. Barnyard grass, (Echinochloa crus-galli) was planted in these pans on a monthly basis, for a total of three months. Efficacy was determined by the number of plants that germinated.

Trifluralin corn field trials were conducted by Midwest Research. Field applications were made with light incorporation to a depth of 1 inch using application rates of 0.55, 1.1, and 2.2 kilograms of active ingredient per hectare. Corn phytotoxicity ratings were taken at 30, 60 and 90 days. The 30 day rating measured corn germination by the number of plants germinated per 100 feet of row. The 60 and 90 day ratings were visual rating of corn injury reported in percent. Weed control ratings were also done visually in percent.

Trifluralin cotton field trials were conducted at Research For Hire, Inc. The soil was sandy loam with a pH of 7.9 and 0.6% organic matter. Then plots were sprayed and mechanically incorporated. The plots were planted with cotton after 48 hours.

RESULTS

Diazinon Microcapsules and Granules

Microcapsules with a 30°C T_M and two different release rates were tested using the standard laboratory bioassay. Figure 2 shows the typical release rates of these two capsule formulations. The fast-release microcapsule formulation has a release rate that is twice as fast as the slow-release formulation at 20°C and four times faster than

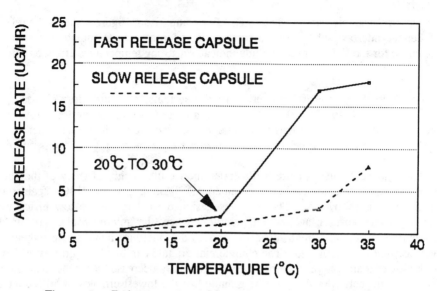

Figure 2 -- Release rates of fast and slow release diazinon
microcapsules in water vs temperature.

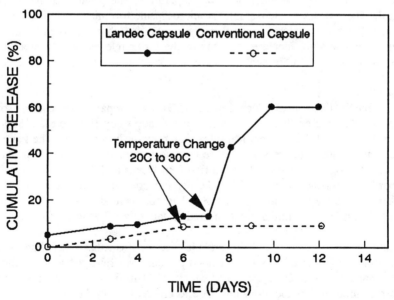

Figure 3 -- Release of insecticide from Landec
and conventional microcapsules

the slow-release microcapsules at 30°C. For comparison, Figure 3 shows the release of Landec microcapsules versus a commercial microcapsule formulation. Clearly, the curve for the Landec microcapsule indicates a strong temperature effect on release rate.

To verify the "on - off" properties of the systems, a granule formulation was cycled from 10°C to 35°C. Figure 4 shows a plot of the microgram/hour release in the "off" and the "on" positions as the granules were cycled. The figure indicates that the active ingredient can be released at will by simply increasing the temperature.

Greenhouse Trials: Figure 5 illustrates the results of the efficacy of the above experimental formulations and a commercial diazinon granule (14G) against the northern corn rootworm. The test was designed to demonstrate low efficacy of Landec microcapsules at low temperature, when the microcapsules are in the crystalline state, and high efficacy above the melting temperature when the microcapsules are activated. The commercial granule exhibited the greatest efficacy at the lower temperature, then pest control gradually decreased after the temperature increase. In fact, after 8 weeks the granule had the lowest efficacy at 50% control. The slow release microcapsule formulation gave little biological control at 20°C. However, the biological control for this formulation increased to 90% when the temperature was increased to 30°C. This higher level of control continued for four weeks at the elevated temperature after which the experiment was terminated. At 2.5 ppm, the slow release formulation was superior to the granule in its duration of control. The efficacy of the fast release microcapsule formulation at 20°C was greater than the slow release microcapsule formulation, but less than the commercial granule. After the temperature increase, efficacy of the faster release system increased from 30% to 90%, where it continued for the next four weeks until the test was terminated.

A Landec diazinon granule (Landec 15G) was prepared for comparison with Diazinon 14G in a northern corn rootworm efficacy study (Figure 6). The test was performed similarly to the microcapsule study previously reported. The Landec 15G gave similar control to the slow release capsule formulation (Figure 5) in that there was little biological control during the first four weeks at 20°C. However, when the temperature was increased to 30°C, the percent control increased to 100% during week 6 and above 90% during week 8. The commercial granule also behaved as expected; high control initially with a gradual tapering off at week 8.

Both the microcapsule and the granule tests confirm that the Landec polymers protect the active ingredients for extended periods then release the active ingredient on thermal demand to obtain the biological effect. This data also confirms that release rates measured in the laboratory correlate to effects seen in the soil.

Trifluralin

Microcapsule formulations with slow and fast release profiles were prepared using a polymer with a 35°C melting temperature. A laboratory bioassay was completed in

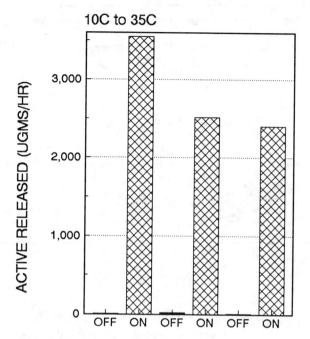

Figure 4 -- Diazinon granule release vs temperature cycling

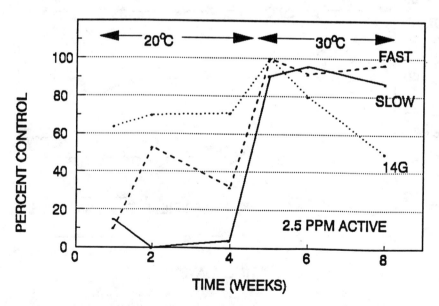

Figure 5 -- Diazinon capsule northern corn rootworm efficacy
at 20°C (4 weeks) and 30°C (4 weeks).

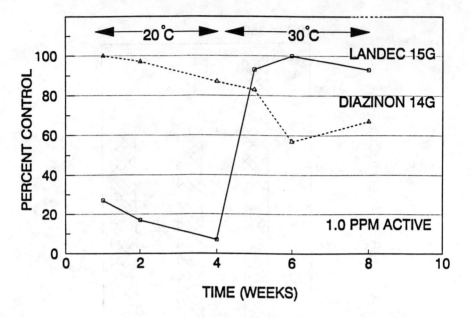

FIG. 6 -- Diazinon granule CRW efficacy
at 20°C (4 weeks) and 30°C (4 weeks).

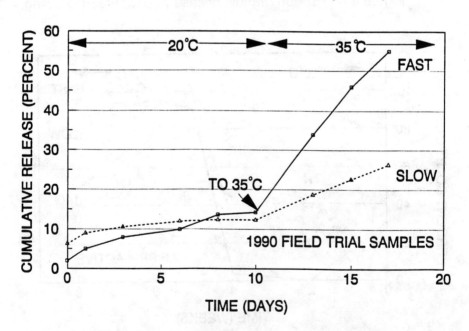

FIG. 7 - Cumulative release of trifluralin capsules
in ethanol:water at 20°C and 35°C.

an ethanol:water solution as described in the methods section of this paper. A typical release profile can be seen in Figure 7. Each formulation's release solution was held at 20°C for 10 days (which is below the melting temperature of the polymer), then moved to 35°C. As the figure shows, one sample was formulated to have more than two times faster release rate above the melting temperature of the polymer. Although the figure clearly indicates that Intelimer can meter the release of an active ingredient based on temperature, it was necessary to evaluate microcapsule formulations in the greenhouse and the field.

Greenhouse Trials: A trifluralin capsule suspension (CS) was prepared with a 30°C switch temperature and evaluated along with a commercial trifluralin emulsifiable concentrate (EC) in the greenhouse with barnyard grass at 30°C to determine if a target control of 90 days could be achieved. Figure 8 shows the efficacy of each formulation after 30 days. The EC gave 100% control at all rates tested, while the CS gave a high degree of control at 0.83 and 1.6 kg AI/ha and no control at 0.42 and 0.21 kg AI/ha. This test indicates that in order to obtain 30 days of control of barnyard grass with Treflan EC, a rate of 0.21 kilograms of active ingredient per hectare is sufficient. This data suggests that Landec microcapsules should be formulated with a similar release to obtain control of this weed species.

Figure 9 compares the efficacy of the EC and the CS formulations with 30°C switch temperatures 90 days after application. At all rates tested, the microcapsule formulation gave superior control over the EC. The CS applied at 0.83 kg AI/ha gave better control than the EC applied at 1.6 kg AI/ha. This demonstrates the ability of the polymer to meter the herbicide for a period of at least 90 days.

Field Trials: Treflan 4EC, Dual™8EC and an experimental Intelimer CS were evaluated in the field to determine the corn phytotoxicity of the formulations. Figure 10 shows the results of this test. Dual (metolachlor) was used as the commercial standard and had no corn phytotoxicity at its normal use rate. All three formulations, at the rates tested, gave equivalent weed control at 30 days. Neither the EC nor the CS significantly prevented the corn from germinating at the 0.55 kg AI/ha rate. However, at the 1.1 kg AI/ha rate, the EC was quite phytotoxic to the corn. The CS allowed greater than 80% germination. This data clearly illustrates the ability of the Intelimer polymers to safen the active ingredient, particularly in crops where phytotoxicity has been a problem in the past.

Figure 11 summarizes the cotton field trial for the control of pigweed at 30, 60 and 90 days intervals. At 30 days , all three formulations gave equivalent control. At 60 days, the EC at 0.55 kg AI/ha was beginning to lose efficacy compared to the EC at 1.1 kg AI/ha and the CS. At 90 days, the EC at 0.55 kg AI/ha exhibited a significant loss of weed control while the CS at 0.55 kg AI/ha was providing equivalent control to that of the EC at 1.1 kg AI/ha. This data illustrates quite clearly that CS formulations can reduce the application rate of trifluralin under field conditions.

LOCATION: PORTERVILLE, CA.

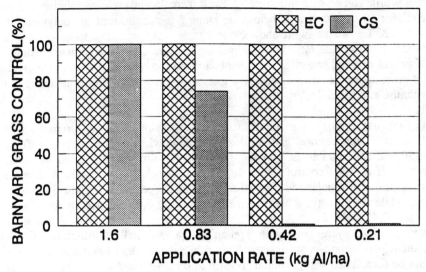

FIG. 8 –   Trifluralin control of barnyard grass in
the greenhouse at 30°C, 30 days after treatment.

LOCATION: PORTERVILLE, CA

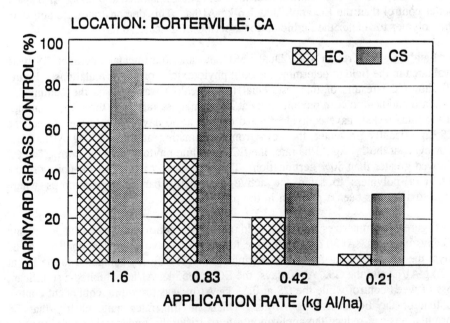

FIG. 9 –   Trifluralin control of barnyard grass
in the greenhouse at 30°C, 90 days after treatment.

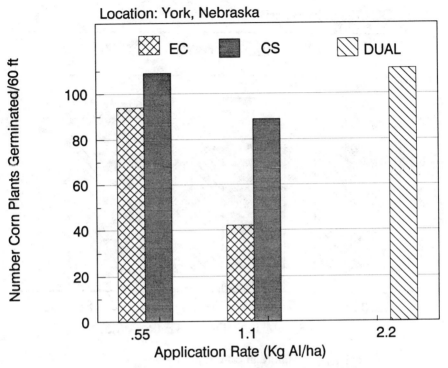

Figure 10 -- Corn phytotoxicity of Intelimer
encapsulated trifluralin 30 days after planting.

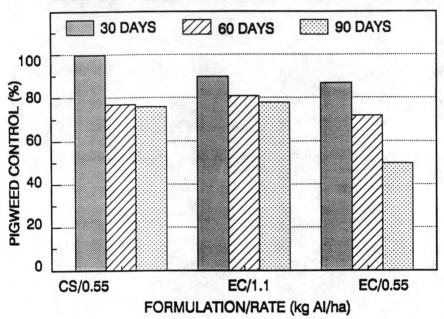

FIG.1 1-- Trifluralin cotton field trial
30,60 and 90 days after treatment.

CONCLUSIONS

Greenhouse and field trials with Intelimer microcapsules of both diazinon and trifluralin gave the same general results:

1. Temperature-activated, delayed release of pesticides was observed in the laboratory.
2. Relative release rates that were observed in vitro were observed in the soil.
3. Efficacy can be achieved using less active ingredient per acre for Intelimer microcapsules than with conventional delivery systems.
4. Crop phytotoxicity can be significantly reduced compared with conventional products.

ACKNOWLEDGMENTS

This work was supported in part by SBIR grant No.88-39410-4710 from the United States Department of Agriculture.   Landec would like to gratefully acknowledge the assistance of Dow Elanco, Griffin Corporation and Ciba Geigy, Ltd. in providing the active ingredients. Laboratory space and equipment was graciously supplied by Raychem Corporation. Without the dedication and hard work of the following individuals, this project would not have been a success: Myrian Villegas and Ray Stewart.

REFERENCES

[1]   Kearney, P. C., "A Challenge for Controlled Release Pesticide Technology", in Controlled Release Pesticides, ACS Symposium Series 53, H.B. Scher (Ed.), The American Chemical Society, Washington D.C., 1977, pp. 30-36.

[2]   Kydonieus, A. F., Controlled Release Technologies: Methods, Theory and Applications, Volume 1, CRC Press Inc., Boca Raton, Florida, 1980.

[3]   Apple, J. W., Walgenbach, E. T., and Knee, W. J.,"Thermal Requirements for Northern Corn Rootworm Egg Hatch", Journal of Economic Entomology, No. 64, 1971, pp. 853-862.

[4]   Ruppel, R. F., "Model for Effective Timing of an Insecticide", Journal of Economic Entomology, No. 77, 1984, pp. 1084-1090.

[5]   Stewart, R. F., U. S. Patent 4,830,855, 1989.

Characteristics of Formulations
and Adjuvants

Kolazi S. Narayanan[1] and Ratan K. Chaudhuri[1]

EMULSIFIABLE CONCENTRATE FORMULATIONS FOR MULTIPLE ACTIVE
INGREDIENTS USING N-ALKYLPYRROLIDONES.

REFERENCE: Narayanan, K. S. and Chaudhuri, R. K., "Emulsifiable
Concentrate Formulations for Multiple Active Ingredients
Using N-Alkylpyrrolidones," Pesticide Formulations and Application
Systems: 11th Volume, ASTM STP 1112, Loren E. Bode and David G.
Chasin, Eds., American Society for Testing and Materials,
Philadelphia, 1992.

ABSTRACT: Higher N-alkyl pyrrolidones, a new and unique class of surface
active solvents have been shown to be excellent for formulating a wide variety of agricultural active ingredients (a.i.) as emulsifiable concentrates
(ECs). The proprietary solvent system we have developed for ECs consists of
four components:

1. Water-soluble lower N-alkyl pyrrolidones, such as N-methyl pyrrolidone
(AgsolEx™1), that are needed for high a.i. loading in the formulations.
AgsolEx™1 is polar, has very high solvency and is biodegradable.

2. Water-insoluble higher N-alkyl pyrrolidones, such as N-octyl pyrrolidone
(AgsolEx™8) and N-dodecyl pyrrolidone (AgsolEx™12) that eliminate crystal
formation and stabilize the EC on dilution with water. These alkyl pyrrolidones are moderately polar, have surfactant and wetting properties, have
low toxicity and are biodegradable.

3. Hydrophobic solvents, such as aromatic petroleum oils and long chain esters, help to improve the quality of the emulsion on dilution with water.

4. Nonylphenol ethyoxylated phosphate ester surfactants.

The active ingredients evaluated in our EC system included: atrazine, bendio-
carb, carbaryl, dichlofluanid, diuron, metolachlor, pendimethalin,
prodiamine, thidiazuron and triforine. Triangular co-ordinate plots were
used to optimize the ternary solvent composition for a fixed weight ratio of
active ingredients and surfactants. By using this solvent optimization technique, a variety of active ingredients having different biological spectra
have been combined into a single EC formulation.

A working model is proposed that explains the universality of the system
and its high stability on dilution with water.

KEY WORDS: Pesticides, formulations, emulsifiable concentrates, N-alkyl pyrrolidones, solvent optimization, surfactants, mixed active ingredients, com-
patibilization

[1]GAF Chemicals Corporation, 1361 Alps Road, Wayne, NJ 07470

Introduction

Solvents that are currently used in the formulation of pesticides are of great concern for both their toxicity and their safety relative to such factors as low flash points and high vapor pressures. Formulators are constantly looking for newer solvents with additional benefits [1].

Emulsifiable concentrates typically consist of active ingredients (a.i.'s) at the highest possible loading in solvents and surfactant(s) at levels required to give maximum stability with the formation of a homogenous emulsion on dilution. Other adjuvants such as stabilizers and rheology modifiers are often optional, but the emulsion produced on dilution to use concentration needs to be reasonably stable in the spray tank, usually for 4-24 hours.

A formulator typically chooses the best solvent system for the a.i. (taking into consideration the cost, toxicity and environmental impact) and emulsifying agents that optimize the desired results. The optimization is therefore done by manipulating the surfactant(s), usually a mixture of non-ionics and anionics [2,3]. Our new system, however, can be optimized for each a.i. by changing the ratios between AgsolEx®'s brand of alkyl pyrrolidone solvents and cosolvents, none of which are on the Environmental Protection Agency's (EPA's) list 1 or 2 [4,5].

Approach

The properties of the alkyl pyrrolidones change considerably as the number of C atoms are increased in the N-alkyl chain [6]. Molecules become surface active when the number of C atoms become ≥ 6. Figure 1 shows the surface tension reduction achieved at various concentrations.

Solvent properties of alkyl pyrrolidones are often characterized in terms of their Hansen's solubility parameters [7,8]. Figure 2 (see also footnote) shows the dispersive, polar and H-bonding parameters relative to the alkyl chain length. It is clear that only the polar and H-bonding components decrease as the alkyl chain is extended.

Solvents can be compared by plotting their fractional solubility parameters in a triangular chart (Figure 3). The fractional parameters for the common solvents were taken from reported values [9], and calculated for the alkyl pyrrolidones. Figure 3 shows the wide latitude of alkyl pyrrolidones (Points 1, 2, 3 and 4) with respect to their polar and dispersive fractional parameters.

Three out of four different solvents (designated by 1, 2 or 3 and K in Fig. 3) having widely different dispersive and polar fractional components were chosen. The choice of solvents is not restricted to three. By mixing these solvents in proper proportions an almost infinite number of solvent types can be made. The proportion of the component solvents and their volume fractions determine the set of fractional solubility parameters for a particular mixture. Thus, with judicious selection of the solvent mixture it is possible to select a system that both dissolves the maximum quantity of a.i. while maintaining emulsion stability on dilution. Other scales for solubility parameters and solvent strength are also available [10].

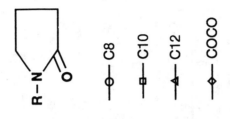

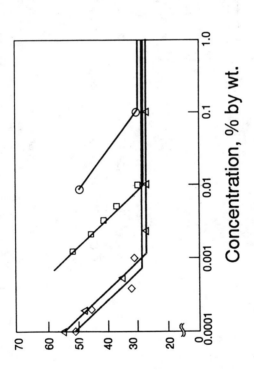

FIG. 1 — Surface tension of aqueous solutions of N-alkylpyrrolidones

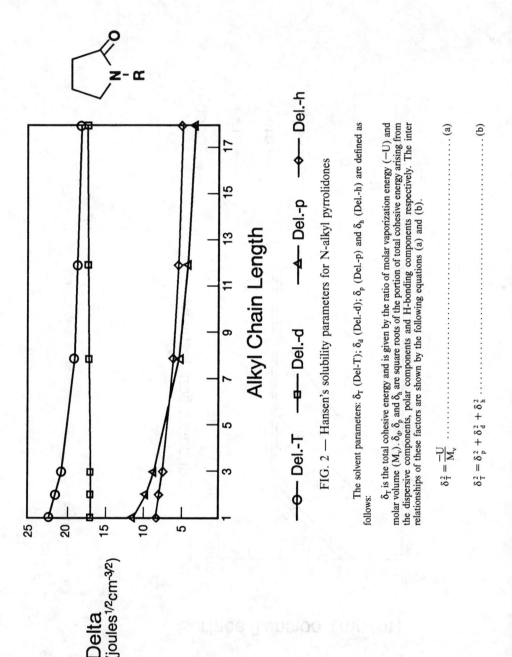

FIG. 2 — Hansen's solubility parameters for N-alkyl pyrrolidones

The solvent parameters: δ_T (Del.-T); δ_d (Del.-d); δ_p (Del.-p) and δ_h (Del.-h) are defined as follows:

δ_T is the total cohesive energy and is given by the ratio of molar vaporization energy ($-U$) and molar volume (M_v). δ_d, δ_p and δ_h are square roots of the dispersive components, polar components and H-bonding components respectively. The inter relationships of these factors are shown by the following equations (a) and (b).

$$\delta_T^2 = \frac{-U}{M_v} \quad \dots \dots \dots \dots \dots \dots \dots \dots \dots \dots \text{(a)}$$

$$\delta_T^2 = \delta_p^2 + \delta_d^2 + \delta_h^2 \quad \dots \dots \dots \dots \dots \dots \text{(b)}$$

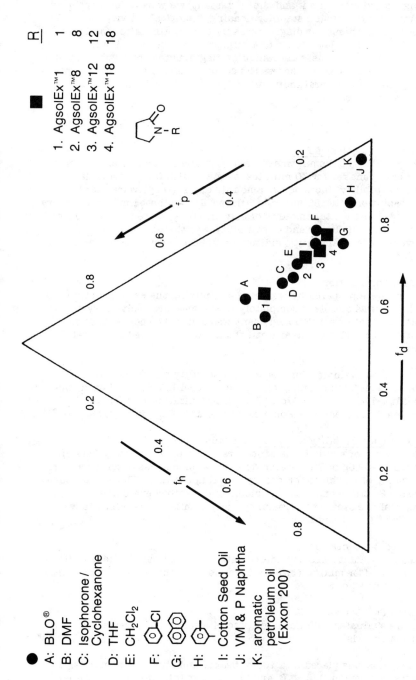

FIG. 3 — Hansen's fractional solubility parameters for N-alkyl pyrrolidones and common solvents

The fractional solubility parameters as shown here are defined as:

$$f_p = \delta_p/\delta_T; \; f_d = \delta_d/\delta_T; \; \text{and} \; f_h = \delta_h/\delta_T.$$

Defined in this form produces better spread for different solvents.

By keeping a fixed ratio of a.i. and the surfactant, we were able to optimize the solvent mixture for emulsion stability, loading, biological activity and other desirable properties. A triangular diagram was used for optimization, and was proven successful for formulating EC's of a number of active ingredients [11]. Others have shown the usefulness of triangular diagrams for optimizing mixed surfactant systems. [2,3] Figure 4 shows the compositions of a number of optimized solvent systems for several individual a.i.'s [12]. This data forms the basis for formulating multiple a.i.'s.

Experimental Section

Determination of Solubilities of A.I.
A weighed quantity of the a.i. was stirred with 10g of the chosen solvent in an automatic rocking shaker for 30 minutes, starting with 0.1-1g of a.i., depending upon its solubility. Incremental amounts of a.i. (0.1g) were added until there was no dissolution after 30 minutes stirring. The highest limit of solubility was thus obtained. Incremental amounts of solvent (0.1g) were then added until a solution was formed; the lower end of solubility was thereby obtained. All determinations were made at ambient conditions ~20-25°C. Results from these tests are shown in Table 1.

Preparation of EC Formulations:
Formulations were prepared by weighing and mixing the exact proportion of ingredients. The a.i. was dissolved completely in the measured solvent system to which the wetting agent or emulsifying agent was added. The contents were mixed in an automatic rocking shaker until the a.i. was dissolved completely, typically 30 minutes.

The samples were evaluated for freeze-thaw stability on storage, ease of emulsification, and emulsion stability on dilution. For dilution, 2g of each concentrate were diluted to 50g using World Health Organization (WHO) standard hard water having a hardness of 342ppm expressed as $CaCO_3$ equivalent.

Evaluation of Emulsion Stability and Ease of Emulsification
A 0.5-2.5g aliquot of emulsifiable concentrate was pipetted into a Nessler tube filled with 47.5-49.5g of WHO water. The initial bloom was observed at zero time and the quality of the bloom graded by visual appearance. The Nessler tube was then inverted 20 times and both the bloom and stability evaluated by volume or height of the sediment (cream/ppt/oil). Similar measurements were made at intervals of 1, 2, 4 and 24 hours.

Stability of Diluted Concentrate
The diluted concentrate was considered "stable" if at final dilutions of 0.2 to 1% the composition, after mixing (twenty inversions), exhibited two mm or less cream and no oil in one hour.

Crystal Formation Studies
Samples were evaluated as follows for precipitation of a.i. (crystal growth) over varying time periods.

1) The diluted sample was placed in a 100ml beaker and stirred continuously. Aliquots were removed at 0, 2, 4, 6, and/or 24 hour intervals and examined under 250x magnification using a 40x30 grid on a 2x2cm slide. The number of

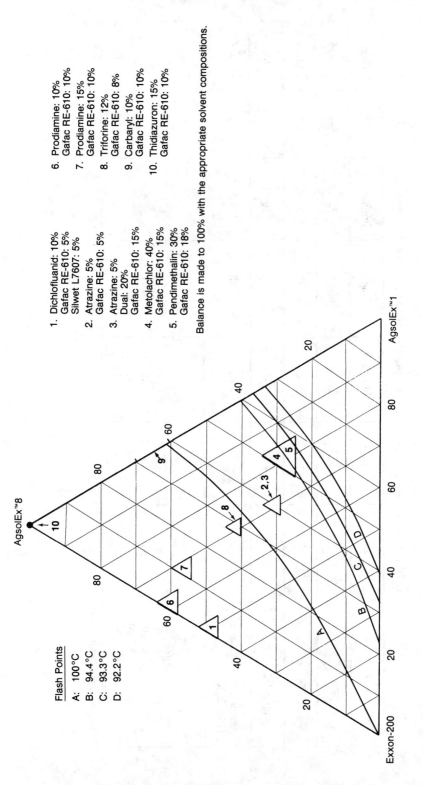

1. Dichlofluanid: 10%
   Gafac RE-610: 5%
   Silwet L7607: 5%
2. Atrazine: 5%
   Gafac RE-610: 5%
3. Atrazine: 5%
   Dual: 20%
   Gafac RE-610: 15%
4. Metolachlor: 40%
   Gafac RE-610: 15%
5. Pendimethalin: 30%
   Gafac RE-610: 18%

6. Prodiamine: 10%
   Gafac RE-610: 10%
7. Prodiamine: 15%
   Gafac RE-610: 10%
8. Triforine: 12%
   Gafac RE-610: 8%
9. Carbaryl: 10%
   Gafac RE-610: 10%
10. Thidiazuron: 15%
    Gafac RE-610: 10%

Balance is made to 100% with the appropriate solvent compositions.

Flash Points
A: 100°C
B: 94.4°C
C: 93.3°C
D: 92.2°C

FIG. 4 — ECs formulated via optimized solvent systems

TABLE 1 — Solubilities of Active Ingredients (% w/w) at Ambient Temperatures

	Atrazine	Thidiazuron	Pendimethalin	Prodiamine	Carbaryl	Triforine
AgsolEx™1	22-23	13-16	66-68	33-42	45-49	34-36
AgsolEx™8	12-15	21-23	42-45	28-32	24-30	13-18
AgsolEx™12	9-10	7-10	—	19-21	16-20	7-10
Aromatic petroleum oil (Exxon 200)	<1	<1	55-58	4-5	1-3	—
	18-20	10-17	56-57			
	10-12	6-8	>57			
	14-15	9-14	56-57			
	16-20	9-15	>57			

AgsolEx™1	AgsolEx™8	Aromatic petroleum oil
50	: 50	: 0
50	: 0	: 50
40	: 30	: 30
55	: 25	: 20

— Denotes, not determined

crystals in ten different grids were counted and averaged. If no crystals were found, second and third aliquots were examined.

2) The remaining portion of the diluted sample was passed through US Standard screens (60, 100, and 250 mesh) and the sediment retained was recorded.

3) The diluted sample was allowed to stand without stirring for 24 hours and then inverted twenty times. An aliquot was then examined for crystals. The remaining portion was passed through screens and retained sediment recorded.

Freeze-thaw Stability

The concentrates were stored for 24 hours from $-10°C$ to $5°C$, thawed to room temperature and then heated to $55°C$ for 24 hours. The alternate cold and hot condition was repeated for three cycles with any separation during the storage recorded. A concentrate is considered "stable" if there is no substantial separation after three cycles.

Results

The high solvent power of N-alkyl pyrrolidones for active ingredients from totally different chemical families is evident from the solubility data shown in Table 1.

Before optimizing ECs of mixed actives, it is important to pinpoint the best system for each component separately. Fig. 4 presents the optimized solvent system for a variety of active ingredients, and Table 2 shows examples of several ECs.

ECs for Atrazine

Because atrazine has extremely low solubility in most solvents, it was chosen as one of the target actives for mixed a.i. formulations. The best solvent system for atrazine was found to be:

AgsolEx™1 = 40%
AgsolEx™8 = 30%
Exxon 200 = 30%

with 5% loading and 5% Gafac RE-610 as the emulsifier. (See Appendix 1 for trade names). The best solvent system was determined as follows.

Eleven different EC's for atrazine were prepared at 5% loading, 5% Gafac RE-610 and the solvent system making up the remainder. The compositions of the 11 solvent systems tested and the results of their stability upon dilution are plotted in a 3 component diagram shown in Figure 5. Shaded figures indicate separation.

Only a narrow solvent range marked by the clear circle was stable when diluted. Although EC's from most solvent systems produced good emulsion (circles), only the optimum system had no crystal separation after 24 h either on standing or on stirring. Tables 3, 4 & 5 summarize the compositions and test results obtained with the best solvent systems.

TABLE 2 — Emulsifiable Concentrates – Prototype Formulations

Active Ingredient	Atrazine 5.0%	Atrazine 5.0% Metolachlor 20.0%	Diuron 5.0%	Pendimethalin 30.0%
Solvent System	AgsolEx™1 36.0% AgsolEx™8 27.0% Exxon Aromatic 200 27.0%	AgsolEx™1 24.0% AgsolEx™8 18.0% Exxon Aromatic 200 18.0%	AgsolEx™8 62.0% Exxon Aromatic 100 24.0%	AgsolEx™1 29.3% AgsolEx™8 13.3% Exxon Aromatic 200 10.6%
Emulsifiers	Gafac RE-610* 5.0%	Gafac RE-610* 15.0%	Gafac RM-710* 9.0%	Gafac RE-610* 16.8%
Flash Point (°C)	> 93.3	> 93.3	> 65.6	> 93.3

Active Ingredient	Metolachlor 40.0%	Prodiamine 15.0%	Prodiamine 15.0%	Thidiazuron 15.0%
Solvents	AgsolEx™1 19.3% AgsolEx™8 8.7% Exxon Aromatic 200 7.0%	AgsolEx™1 9.75% AgsolEx™8 40.50% Exxon Aromatic 200 24.75%	AgsolEx™1 9.75% AgsolEx™8 20.25% AgsolEx™12 20.25% Exxon Aromatic 200 24.75%	AgsolEx™8 30.0% AgsolEx™12 45.0%
Emulsifiers	Gafac RE-610* 25%	Gafac RE-610* 10%	Gafac RE-610* 10%	Gafac RE-610* 10%
Flash Point (°C)	> 93.3	> 93.3	> 93.3	> 93.3

TABLE 2 — Emulsifiable Concentrates – Prototype Formulations (continued)

Active Ingredient	Bendiocarb 15.5%	Euparen 10.0%	Carbaryl 10.0%	Triforine 10.0%
Solvent System	AgsolEx™1 15.9% AgsolEx™8 29.4% Exxon Aromatic 100 32.0%	AgsolEx™8 40.0% Dibutyladipate 40.0%	AgsolEx™1 16.0% AgsolEx™8 48.0% AgsolEx™12 16.0%	AgsolEx™8 80.0%
Emulsifiers	Gafac RM-710* 7.2%	Gafac RE-610* 5.0% Silwet L-7607** 5.0%	Gafac RE-610* 5.0% Pegol L-31* 5.0%	Gafac RE-610* 5.0% Igepal CO-630* 5.0%
Flash Point (°C)	~65.6	> 93.3	> 93.3	> 93.3

*Rhone Poulenc
**Union Carbide

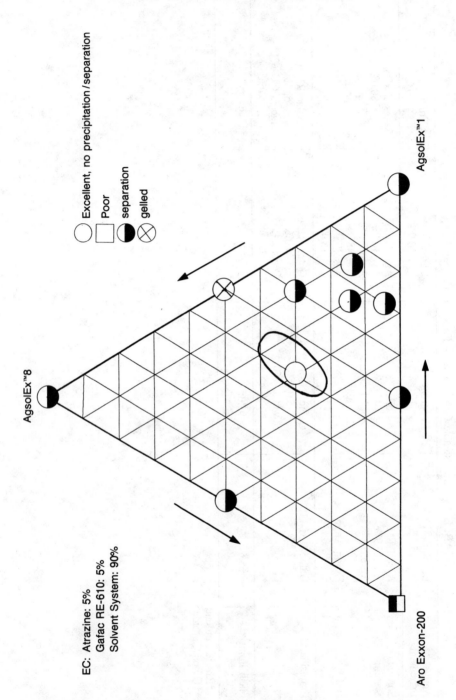

FIG. 5 — Optimization of solvent system for Atrazine.

TABLE 3 — Compositions of Solvent Optimized Emulsifiable Concentrates

Ingredients	Compositions, Weight %							
	1	2	3	4	5	6	7	8
AgsolEx™1	36.0	19.25	24.0	28.0	32.0	32.0	30.4	25.7
AgsolEx™8	27.0	8.75	18.0	21.0	32.0	32.0	30.4	43.7
Aro Exxon 200	27.0	7.00	18.0	21.0	16.0	16.0	15.2	8.6
Gafac RE 610	5.0	25.00	15.0	10.0	10.0	8.0	12.0	10.0
Atrazine Tech 97%	5.0	—	5.0	5.0	4.0	4.0	4.0	—
Metolachlor Tech	—	40.00	20.0	15.0	—	—	—	—
Prodiamine Tech 94.7%	—	—	—	—	6.0	8.0	8.0	—
Carbaryl Tech 99%	—	—	—	—	—	—	—	7.0
Triforine Tech	—	—	—	—	—	—	—	5.0
Total	100.0	100.00	100.0	100.0	100.0	100.0	100.0	100.0
Optimized, 3-Component Solvent System								
AgsolEx™1	40	55	40	40	40	40	40	33
AgsolEx™8	30	25	30	30	40	40	40	56
Aro Exxon 200	30	20	30	30	20	20	20	11
Total	100	100	100	100	100	100	100	100

Formulations 1 and 2 contain single active ingredients. Formulations 3 through 8 contain two active ingredients. Solvent systems for the combined actives were derived from optimized solvent systems for the individual actives. Several EC's were screened. This Table shows only those EC's that passed the screening evaluations — see Tables 4 and 5.

TABLE 4 — Results of Emulsion Stability on Standing (Nesslers Tube)

EC Used[1]	1		2		3		4		5		6		7		8	
Dilution	2.5g of the EC / 50g with "342 ppm hardness" water															
Bloom[2] 0 Time	Excellent		Excellent		Excellent		Excellent		Excellent		Excellent		Excellent		Good	
After 20 Inversions	Excellent		Excellent		Excellent		Excellent		Excellent		Excellent		Excellent		Excellent	
Solids/Cream/Oil, mm³	T	B	T	B	T	B	T	B	T	B	T	B	T	B	T	B
Time: 0 h	0	0	0	0	0	0	0	0	0	0	0	0	0	0	0	0
1 h	0	0	0	0	0	0	0	0	0	TR	0	0	0	1	0	TR
2 h	0	0	0	0	0	0	0	0	0	1	0	2	0	2	0	1
4 h	4	0	0	0	0	0	0	0	0	1	0	3	0	2	0	1
24 h	4	0	0	0	0	0	0	0	0	3	0	3	0	2	0	3
Crystals at ×250[4]	0		0		*		0		*		**		*		0	
Filtration Through Screens After 24 h Standing and 20 Inversions																
60 mesh	—		—		—		—		—		—		—		—	
100 mesh	—		—		—		—		—		—		—		—	
250 mesh					+		+									

1 See Table 4 for a listing of ingredients in these numerically cross-referenced ECs.
2 Bloom Excellent = thick emulsion cloud with no separation
  Good = Emulsion cloud may be thin or may exhibit trailing. Small number of oil droplets within cloud.
  Poor = Many oil droplets in cloud, some droplets may separate from cloud.
3 T: Top; B: Bottom; — means no sediments; + means more than a trace, <1%. TR means trace.
4 0 = no crystals; * = 0-50 crystals; ** = 50-100 crystals.

TABLE 5 — Microscopic Observation of Crystal Growth from Formulations on Dilution and Stirring[2]

EC	1	2	3	4	5@	6@	7@	8@
Dilution			2.5g of the EC / 50g with "342 ppm hardness" water					
Time: 0 h	0	0	0	0	0/*	0/*	0/*	0/*
2 h	0	0	0	0	0	0	0	0
4 h	0	0	0	0	0	0	0	0
6 h	0	0	0	0				
24 h	0	0	*	0	0	0	0	0
After 24 h, filtered through screens								
60 mesh	—	—	—	—	—	—	—	—
100 mesh	—	—	TR	—	—	—	—	—
250 mesh			—	—	—	—	—	—

1 See Table 4 for a listing of ingredients in these numerically cross-referenced ECs.

2 0 = no crystals; * = 1-10 crystals (crystal dimension 1-40 μ).

@ There was no change from 0-24 h; a few tiny crystals (1-5) per view were occasionally seen.

ECs for Metolachlor

The optimum solvent systems were determined for EC's of 40% metolachlor using 15% Gafac RE-610 as the surfactant. The optimized solvent compositions are shown by the loop enclosing the clear circles in Figure 6. Shaded circles indicate separation.

ECs with atrazine/metolachlor

By superimposing Figures 5 and 6, the overlapping optimized region could be determined as in Figure 7. EC's containing 20% & 25% combined actives with atrazine = 5% and metolachlor 15%/20% were prepared. The Gafac RE-610 surfactant concentration was kept at 15-20%. Several solvent compositions around the overlapping "optimized area" were used to make up the 100% material balance.

Compositions and performance of the best EC system for atrazine/metolachlor are shown in Tables 3, 4, and 5. It is clear that the best solvent composition is within the predicted area. This approach can be used as a general tool for compatibilizing different a.i.'s in the ECs.

ECs with atrazine/Prodiamine

The optimized solvent composition for prodiamine was found to be a loop around "area 7" in figure 4 with 15% loading and 10% Gafac RE-610. Optimized solvent compositions for atrazine are given by "area 2" in Figure 4 or the loop in Figure 5. Since these do not overlap, the optimum compositions for the combined a.i.'s are between the two optimized areas.

As a test case the following solvent composition was screened:

AgsolEx™1 = 40%
AgsolEx™8 = 40%
Exxon 200 = 20%

with 8-10% prodiamine, 5% atrazine, and 8-10% Gafac RE-610. The solvent composition made up the balance to 100%. Tables 3, 4 and 5 show the EC compositions and test results. It is clear then that ECs can be made using our concept, and that compositions falling on the line between center points in '2' and '7' (in Figure 4) should perform better than other mixtures.

ECs with Triforine/Carbaryl

Similarly, compositions on the line joining the optimized area for the individual a.i.'s (areas '8' and '9' in Figure 4) was used as a test case. An EC was made with 7% carbaryl, 5% triforine and 10% Gafac RE-610 with the balance being the solvent composition shown below, a composition that is colinear with points '8' & '9' in Figure 4.

AgsolEx™1: 33%
AgsolEx™8: 56%
Exxon 200: 11%

The composition and results of screening are shown in Tables 3, 4 & 5. It is clear that an EC of carbaryl and triforine is achievable and passed the 24 h stability test.

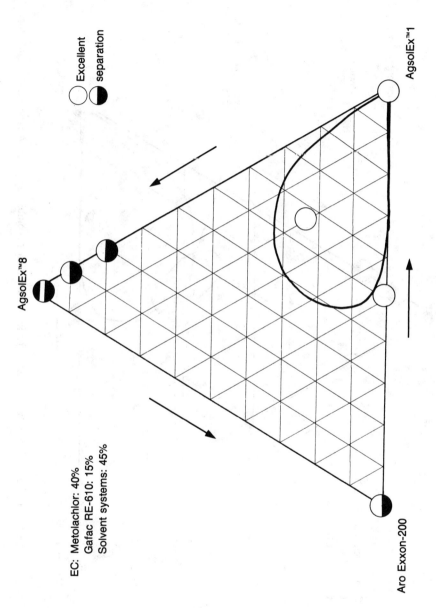

FIG. 6 — Optimization of solvent system for Metolachlor.

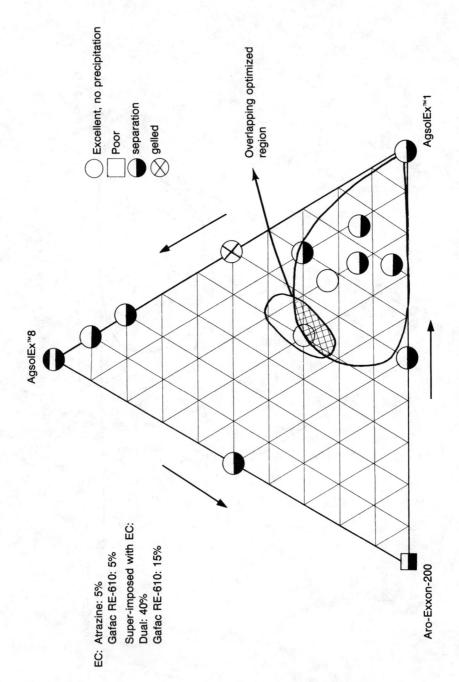

FIG. 7 — Superimposed optimized solvent systems for EC formulation of combined actives: Atrazine and Metolachlor

Discussion

The examples illustrated herein show the tremendous potential for the use of N-alkyl pyrrolidones in EC systems. We have chosen active ingredients that are difficult to formulate as ECs by conventional solvents to demonstrate the superior properties of N-alkyl pyrrolidones.

A microphotograph of a typical emulsion taken 24 h after dilution from a typical EC in this system had a uniform emulsion droplet size of 5–10μ with no crystals.

A combination of a lower alkyl pyrrolidone (AgsolEx™1) for high solvency and a higher alkyl pyrrolidone (AgsolEx™8) for emulsion stability is needed. In the examples shown in Table 2, two different fractions of aromatic petroleum oil are used as examples of the third solvent in the system.

The three component solvent system comprising AgsolEx™1, AgsolEx™8 and aromatic Exxon 200 has the further advantage of high flash points ($\geq 87.8°C$ for all compositions and $\geq 93.3°C$ for most). Flash point contours are shown in Figure 4. [13].

The third solvent needs only to have a dispersive fractional coefficient of $> 90\%$. Vegetable oils such as soybean oil have been successfully used as the 3rd solvent to formulate EC/ME for various insecticides that normally have poor solubility in soybean oil [11].

Structure of EC Systems

Below is a working hypothesis for the present EC systems. In system 1 all the components are super-solvents which may form the micelles diagramed in Figure 8A. In System 2 one of the components is a poor solvent (water insoluble oil), and a reverse micelle may be produced as shown in Figure 8B.

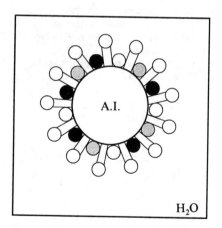

System 1

It is proposed that all solvents form part of the solvation shell. On dilution with water a micelle would form with an interpenetrating triple solvent shell. The AgsosolEx™8/12, complexed with anionic surfactants, would orient on the micellar surface with hydrophilic groups into the outside continuous water phase.

FIG. 8A — Micelle model in water.

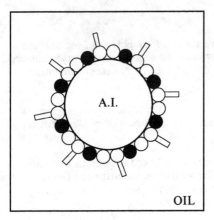

FIG. 8B — Reverse micelle model in oil.

System 2

It is proposed that this system would be a micelle with an interpenetrating double solvent-solvation layer. The third solvent, a water insoluble oil, is the continuous phase. AgsolEx™8/12 complexed with anionic surfactants, would in this instance orient on the micellar surface with the hydrophobic groups towards the outside. On dilution with water, the system would probably invert to a system 1 type.

In the model system the a.i. is solvated and what has to be emulsified is the solvent shell in water. After application the micelle interacts with the lipid layer of the target, the micellar structure breaks and the a.i. is free to act on the target. Thus, based on this hypothesis, we would expect a well-directed, target-specific formulation system.

Emulsifiers

Several emulsifers were screened initially. Nonylphenol ethoxolated phosphate esters, Gafac RE-610, and Gafac RM-710 were shown to produce excellent bloom and emulsion stability. Other less effective emulsifiers were Igepal CO-730 and Ninate 401. [14]

The effectiveness of the anionic phosphate esters in conjunction with AgsolEx™8/12 could be due to complexation involving protonation of the amide carbonyl and formation of an ion-pair involving the N atom of the pyrrolidone ring and the anion of the surfactant. [Fig. 9] This would give rise to a larger area of hydrophobic interaction. Synergysm between N-octyl pyrrolidones and LAS have been shown and explained by such a mechanism. [15,16]

FIG. 9 — Complexation of AgsolEx™ solvents with anionic surfactant

Optimization with respect to all components

Examples illustrated here are only lead formulations showing optimized solvent systems with respect to the chosen surfactant and its level. The best solvent system for solubilizing the a.i. is not necessarily the best solvent system for the EC system. For example AgsolEx™1 alone is a super solvent for many a.i.'s. but the ECs are not always stable on dilution. A 3-component solvent system is desirable in most cases. Conversely, a solvent system has to have a high solubility for the active -- usually about 1.2 to 1.5 times more solubility than the desired loading in the final EC.

The three component solvent system can be further optimized with the correct combination of surfactants. An optimization can then be accomplished in terms of biological activity, emulsion stability, and cost. The following offers an experimental design for optimization [17] in terms of the 3-component solvent systems and surfactants for maximum a.i. loading and stability.

The results can be treated semiquantitatively using polynomial fits for the desirability functions as shown in eq (1) in terms of the weight fraction of the 'n' components: (χ_i). Where Φ is the desirability function which is a product function of each desired individual response (Φ_i) derived by observations. (eq 3)

$$
\begin{aligned}
\Phi = {} & \sum^n \beta_i \, \chi_i \\
& + \sum^n \beta_{ij} \chi_i \chi_j \\
& + \sum^n \beta_{ijk} \chi_i \chi_j \chi_k \\
& + \sum^n \beta_{ijkl} \chi_i \chi_j \chi_k \chi_l \cdots + \cdots\cdots + \cdots\cdots\cdots \\
& + \sum^n \beta_{ijklmn} \chi_i \chi_j \chi_k \chi_l \chi_m \chi_n \quad\cdots\cdots\cdots\cdots (1)
\end{aligned}
$$

$$\sum \chi_i = \text{constant} \quad\cdots\cdots\cdots\cdots\cdots\cdots (2)$$

$$\Phi = \prod_i \Phi_i \quad\cdots\cdots\cdots\cdots\cdots\cdots (3)$$

β's are the coefficients of interactions

Since increasing the number of components makes the calculation more complicated, it is desirable to run optimization designs in terms of a 3-component solvent composition for each fixed weight ratio of surfactants. Under these conditions equation (1) reduces to the simpler form shown below.

$$
\begin{aligned}
\Phi = {} & \beta_1 \chi_1 + \beta_2 \chi_2 + \beta_3 \chi_3 \\
& + \beta_{12} \chi_1 \chi_2 + \beta_{13} \chi_1 \chi_3 + \beta_{23} \chi_2 \chi_3 \\
& + \beta_{123} \chi_1 \chi_2 \chi_3 \quad\cdots\cdots\cdots\cdots\cdots (4)
\end{aligned}
$$

$$\text{and } \chi_1 + \chi_2 + \chi_3 = \text{Constant} \quad\cdots\cdots\cdots (5)$$

These calculations give optimization contours that can be drawn in triangular plots. Others have used similar polynominals to optimize wettable powder formulations [18].

Conclusion

This work shows that a combination of N-alkyl pyrrolidones can be used to formulate ECs of active ingredients that are otherwise difficult or impossible to formulate. ECs have been successfully made of a wide variety of a.i.'s belonging to various chemical families with widely different solubility profiles. Using the techniques discussed here, it is possible to formulate mixed a.i.'s having different biological spectra into a single EC. These ECs are environmentally acceptable due to the biodegradability and high flash point of the component solvents.

APPENDIX 1 — Chemical Description of Trademark Chemicals

Trade Name	Supplier	Chemical Name
AgsolEx™1 AgsolEx™8 AgsolEx™12	GAF GAF GAF	N-methyl pyrrolidone N-octylpyrrolidone N-dodecylpyrrolidone
Aromatic Exxon 200	Exxon	Heavy aromatic solvent naphtha (petroleum) consists predominantly of C_9–C_{15} aromatic hydrocarbons bp: 217–293 °C
Aromatic Exxon 100	Exxon	Light aromatic solvent naphtha (petroleum) consists predominantly of C_8–C_{10} aromatic hydrocarbons bp: 152–168 °C
Gafac RE-610	Rhone Poulenc	Poly (oxy-1,2-ethanediyl) $\propto$-(nonylphenyl)-omega-hydroxy-phosphate
Gafac RM-710	Rhone Poulenc	Poly (oxy-1,2-ethanediyl) $\propto$-(dinonylphenyl)-omega-hydroxy-phosphate
Igepal CO-630	Rhone Poulenc	Ethoxylated nonylphenol containing 9 EO units
Pegol-L-31	Rhone Poulenc	Ethoxylated polyoxypropylene MW: 1100
Silwett L-7607	Union Carbide	Polyalkylene oxide modified polydimethyl siloxane
Igepal CO-730	Rhone Poulenc	Ethoxylated nonylphenol containing 15 EO units
Ninate 401	Stepan	Calcium alkyl benzene sulfonate
Atrazine	Ciba Geigy	6-chloro-N-ethyl-N'-(1-methyl ethyl)-1,3,5,triazine-2,4-diamine
Bendiocarb	Noram	2,2-dimethyl-1,3-benzodioxol-4-yl methylcarbamate
Carbaryl	Union Carbide	1-Naphthalenyl methyl carbamate
Diuron	Dupont	N'-(3,4-dichlorophenyl)-N,N-dimethylurea
Euparen (Dichlofluanid)	Bayer	1,1-dichloro-N-[(dimethylamino) sulfonyl]-1-fluoro-N-phenyl methane sulfenamide
Metolachlor	Ciba Geigy	2-chloro-N-(2-ethyl-6-methylphenyl)-N-(2-methoxy-1-methylethyl) acetamide
Pendimethalin	American Cyanamid	N-(1-ethylpropyl)-3,4-dimethyl-2,6-dinitro benzenamine
Prodiamine	Sandoz	2,4-dinitro-N^3,N^3-dipropyl-6-(trifluoromethyl)-1,3-benzene diamine
Thidiazuron	Noram	N-phenyl-N'-1,2,3-thiadiazol-5-ylurea
Triforine	Shell	N,N'-[1,4-piperazine diyl bis (2,2,2-trichloro ethylidene)] bisformamide

References and Notes

[1] Doyal, K.J., McKillip, W.J., Shin, C.C., and Rickard, D.A., "Comparison of the cyclic ether-alcohol tetrahydro-furfuryl alcohol to other known solvents and adjuvants," Second International Symposium on Adjuvants for Agrochemicals, Virginia Polytechnic Institute and State University, Blacksburg, Virginia, USA, Aug. 2, (1989).

[2] Bolts, M. F., "Pesticide Formulations and Application Systems," 8th vol. ASTM, STP-980. D.A. Hovde and G.B. Besstman, editors, ASTM, Philadelphia, (1988), pp. 77-87.

[3] Skelton, P.R. Munk, B.H. and Collins, H.M., "Pesticide Formulations," ibid, (1988), pp. 36-45.

[4] List 1 and List 2 are the lists compiled by EPA classifying compounds considered toxic and compounds considered potentially toxic, respectively. See 40 CFR 154.7 dated April 22, 1987.

[5] Agsol™Ex 8/12 are approved to be used in Europe. US approval is pending.

[6] See GAF brochures on "Ag-Products."

[7] Hansen's Solubility Parameters for N-alkyl pyrrolidones were evaluated from group contributions from published data. See "CRC Handbook of Solubility Parameters," Barton A.F.M., editor, (1983), Published by CRC Press, Ch. 6, pp. 61-89.

[8] Teas, J.P., Journal Paint Tech., (1968) 40, pp. 19-25.

[9] Barton, A.F.M., "CRC Handbook of Solubility Parameters and Other Cohesion Parameters," CRC Press Inc., Florida (1983), pp. 167-190; Use of solvent parameters in pesticide formulations is documented. See Meusberger, K., "Advances in Pesticide Formulation Technology," ACS Symposium Series 254 (1983), Scher, H.B., editor, Ch. 9, pp. 121-137.

[10] Rutan, S.C., et al, J. Chromatogr (1989) 463, pp. 21-37.

[11] Narayanan, K.S., Chaudhuri, R.K., Dahanayake, M., Several Patent applications on the use of N-alkyl pyrrolidones for formulating EC systems are pending. We have successfully formulated dozens of different a.i.'s belonging to totally different chemical species using the 3-component solvent system optimization technique and a fixed weight ratio of a.i. and emulsifying agent. The emulsifying agent used was Gafac RE-610.

[12] Narayanan, K.S., and Chaudhuri, R.K., "Use of N-Alkyl pyrrolidones in emulsifiable concentrate formulations," Seventh International Congress of Pesticide Chemistry, eds. H. Frehse, E. Kesseler-Schmitz and S. Conway, IUPAC, Hamburg, Germany, Aug. (1990)-Vol. II, 05B-20.

[13] Narayanan, K.S., Flash Point Contours for Multi-component Solvent Systems - AgsolEx™1/AgsolEx™8 Petroleum Oil, Unpublished work (1990).

[14] Dahanayake, M., and Newton, M., Unpublished work (1988). Also see Saito, Y., Sato, T., and Anazuva, I., "Effect of Molecular Weight Distribution of Nonionic Surfactants on Stability of O/W Emulsions," Journal of the American Oil Chem. Soc., (1990), 3, p. 145.

[15] Rosen, M.J., et al, Journal of the American Oil Chem. Soc., (1989), 66, p. 998.

[16] Rosen, M.J., et al, ibid, (1988) 4, p. 1273.

[17] Deming, S.N. and Morgen, S.L., "Experimental Design: A Chemometric Approach," Elsvier, Amsterdam, The Netherlands (1987).

[18] Mookerjee, P.K., "Advances in Pesticide Formulation Technology," ACS Symposium Series 254 (1983), Scher, H.B., editor, Ch. 8, pp. 105-119.

Charles A. Duckworth and Kathleen S. Cearnal

THE EFFECT OF CARRIER TEMPERATURE ON PESTICIDE EMULSIFICATION OR DISPERSION CHARACTERISTICS

REFERENCE: Duckworth, C. A. and Cearnal, K. S., "The Effect of Carrier Temperature on Pesticide Emulsification or Dispersion Characteristics," Pesticide Formulations and Application Systems: 11th Volume, ASTM STP 1112, Loren E. Bode and David G. Chasin, Eds., American Society for Testing and Materials, Philadelphia, 1992.

ABSTRACT: The results of a study determining the effect of carrier temperature on the emulsification or dispersion and redispersion of commercial pesticide formulations are detailed. A series of commercially available emulsifiable concentrate, liquid flowable and dry flowable formulations were evaluated in waters of varying hardness and temperature. This study was extended to several model emulsfiable concentrate systems to determine what adjustments in emulsifier type were required to achieve consistent emulsion performance in water varying in temperature.

KEY WORDS: emulsification, redispersion, pesticide, water hardness, temperature.

INTRODUCTION

This paper details the results of a study to determine the effect of carrier temperature on the emulsification or dispersion and redispersion of pesticide formulations. A series of commercially available emulsifiable concentrate (EC), liquid flowable (LF) and dry flowable (DF) formulations were evaluated in waters of varying hardness and temperature.

The current ASTM method, (E1116-86), recommends a temperature of 25°C to test the emulsification properties of EC formulations, and suggests that other water temperatures can be tested depending upon the end use conditions of the pesticide. During the evaluation of commercial pesticide products at Monsanto, significant differences in the emulsion performance of various EC formulations have been observed with only minor deviations

Mr. Duckworth and Ms. Cearnal are research scientists in the Formulation Technology Section of Monsanto Agricultural Co., a unit of Monsanto Company, 800 N. Lindbergh Blvd., St. Louis, MO 63167.

in temperature (<2°C). One objective of this study was to determine how the mixing and dispersion/redispersion characteristics of various commercial EC formulations vary as the carrier temperature is lowered from the 25°C recommended in the test procedure to a temperature closer to that of well water as it is taken from the well.

A second objective of this study was to compare the performance of various commercial liquid flowable and dry flowable formulations at the suggested 25°C test temperature versus a temperature closer to that of well water.   It was suspected that the mixing and dispersion/redispersion characteristics would be negatively impacted by lowering the carrier temperature.   Since no ASTM method has been published governing the evaluation of these formulation types, the EC method was modified and employed.

A final objective of the study was to develop several model EC systems without active ingredient to determine what emulsifier adjustments would be required to achieve emulsion performance at near well water temperatures that is similar to the performance observed at 25°C.   HLB values were determined for the resulting emulsifier blends for comparison purposes.

MIXING AND REDISPERSION OF COMMERCIAL PESTICIDES

<u>Experimental Details</u>

The objective of this experiment was to determine how the mixing and dispersion/redispersion characteristics of various commercial EC, liquid flowable and dry flowable formulations vary as the carrier temperature is lowered from the recommended testing temperature of 25°C to a temperature closer to that of well water as it is taken from the well.

Bloom, dispersion and redispersion evaluations were carried out using ASTM "Standard Test Method for Emulsification Characteristics of Pesticide Emulsifiable Concentrates", E1116-86, except that Nessler tubes were used rather than graduated cylinders.   In the case of the EC formulations 0 and 1500 ppm hardness waters were evaluated in addition to the waters outlined in the method in order to obtain data outside the normal testing range.   The samples were kept at constant temperature by use of a water bath or incubator.   A constant dilution rate of 5 ml/95 ml carrier was used for the EC and liquid flowable evaluations, and 5 g/95 ml carrier for the DF evaluations.   Blooms were not evaluated for the liquid or dry flowables.

The EC and SC formulations were evaluated at carrier temperatures of 25 and 7.2°C, while the DF formulations were evaluated at 25 and 4.4°C.

<u>Experimental Results</u>

<u>Emulsifiable Concentrates:</u> The results of evaluations with various commercially available emulsifiable concentrates is shown in Table 1.   The products were selected due to their availability and the fact that they represent chlorinated, aromatic and solventless systems.   For the purposes

TABLE 1 -- Results for commercial emulsifiable concentrates

Water	77°F					45°F				
	Bloom	Observed separation, ml 15 min.	1 hour	24 hour	Inv. to suspend	Bloom	Observed separation, ml 15 min.	1 hour	24 hour	Inv to suspend
EC-1										
0 ppm	perfect	tr crl	1 crl	1crl/2 cr	3	poor	1 cr	2 crl	5 cr/1 crl	2
35	perfect	tr cr	tr cr	1 cr	2	good	0.5 cr	1.5 cr	5. cr	3
342	perfect	0	0	tr cr	2	perfect	tr cr	0.5 cr	2 cr	3
1000	perfect	0	0	9 cr	2	perfect	tr cr/tr o	tr cr/tr o	1 crl/tr o	4
1500	good	0	tr cr	6 cr	2	perfect	tr crl	tr crl	.5 crl/.5 o	3
EC-2										
0 ppm	perfect	2 crl	3 crl	5 crl	3	poor	3 crl	4 crl	5 crl	>10
35	perfect	1 crl	2 crl/tr o	4 crl/tr o	5	poor	3.5 crl	4 crl/tr o	4 crl/tr o	3
342	perfect	0	tr o	0.5 o	3	good	1.5 crl	3 crl/tr o	4 crl/tr o	3
1000	poor	2.5 crl	4 crl	4.5 crl	5	perfect	tr crl	3 crl	3 crl	3
1500	poor	3 crl	4 crl	4.5 crl	8	perfect	tr cr	1 crl/tr o	4 crl	10
EC-3										
0 ppm	poor	1 cr/ph o	3 cr/ph o	4 crl/ph o	3	poor	2 crl/ph o	3 crl/ph o	5 crl/ph o	4
35	poor	1 cr/0.5 o	2 crl/1 o	3 crl/1 o	5	poor	2 crl/1 o	2 crl/1 o	4 crl/1 o	5
342	poor	1 crl/1 o	2 crl/1 o	3 crl/1 o	4	poor	1 crl/1 o	2 crl/1 o	4 crl/1 o	4
1000	poor	1 crl/1 o	2 crl/1 o	3 crl/1 o	4	poor	1 crl/1 o	2 crl/1 o	4 crl/1 o	4
1500	poor	tr crl/1 o	tr crl/1 o	2 cr/1 o	7	poor	tr crl/1 o	2 crl/tr o	4 cr/1 o	8
EC-4										
0 ppm	good	0.5 cr	2 cr	5 cr	6	poor	1 cr	2 cr	5 cr	>10
35	perfect	tr cr	1 cr	3 cr	4	poor	0.5 cr	2 crl	4 crl	2
342	perfect	tr cr	tr cr	1 crl/tr o	2	poor	tr cr	1 crl	3 crl	2
1000	perfect	0	0	tr cr/ph o	2	perfect	tr crl	tr crl	2 crl/tr o	3
1500	poor	3 crl	4.5 crl	4.5 crl	>10	perfect	0.5 cr	0.5 cr	tr crl/2 o	6
EC-5										
0 ppm	floats	0	0	1 cr	1	floats	0	0	0	0
35	floats	0	0	0	4	floats	0	0	0	1
342	floats	tr cr	2 crl/tr o	4 o	2	floats	0	0	1 crl/tr o	2
1000	floats	2.5 crl	5 crl/tr o	4 o	2	floats	1 crl	4 crl	5 crl	2
1500	floats	2 crl/tr o	tr crl/5 o	5 o	>10	floats	2 crl/tr o	4 crl/tr o	5 crl	3

of this investigation the bloom was rated as "perfect" if a thick emulsion cloud reached the bottom of the Nessler tube without separation of any type, "good" if the bloom was thin or if separation was observed at one inch or less from the bottom of the tube, and "poor" if the emulsion cloud contained oil droplets or oil separation above one inch from the bottom of the tube. All stability readings are as ml separation with: "cr" denoting cream, "crl" denoting creamy oil, "ph" denoting pinheads of oil and "o" denoting oil.

In all cases, a change in performance was noted on decreasing the carrier temperature from 25 to 7.7°C. In general, there appeared to be a shift towards better performance in harder water as the temperature of the water was decreased.

Liquid Flowables: The results of evaluations with various commercially available liquid flowables is shown in Table 2. In general, there was no significant difference observed in mixing or dispersion/redispersion when the carrier temperature was decreased from 25 to 7.7°C. In the case of liquid flowable 2, there appeared to be an increase in flocculation at 7.7°C, which decreased the hard packing, and less inversions were required to resuspend than at 25°C.

TABLE 2 -- Results for commercial liquid flowables

Water	77°F (25°C)				45°F (7.7°C)			
	Observed separation			Inv. to	Observed separation			Inv. to
	15 min	1 hr	24 hr	suspend	15 min	1 hr	24 hr	suspend
LF-1								
35 ppm	trace[a]	trace	1 ml	>10	trace	trace	6 ml	6
342	trace	trace	4 ml	3	trace	trace	4 ml	10
1000	trace	trace	10 ml	3	trace	trace	2 ml	>10
LF-2								
35 ppm	2 ml	4 ml	3 ml	7	10 ml	10 ml	9 ml	4
342	2 ml	4 ml	4 ml	8	5 ml	5 ml	8 ml	4
1000	3.5 ml	6 ml	4 ml	7	3 ml	3 ml	6 ml	4
LF-3								
35ppm	trace	trace	1 ml	>10	trace	trace	1 ml	>10
342	trace	trace	3 ml	>10	trace	trace	1 ml	>10
1000	trace	trace	4 ml	6	trace	trace	1 ml	>10
LF-4								
35ppm	trace	0.5 ml	3 ml	5	trace	trace	2 ml	5
342	trace	0.5 ml	3 ml	5	trace	trace	2 ml	6
1000	trace	0.5 ml	3 ml	5	trace	trace	2 ml	6
LF-5								
35 ppm	trace	trace	1 ml	>10	trace	trace	1 ml	>10
342	trace	trace	1 ml	>10	trace	trace	1 ml	>10
1000	trace	trace	1 ml	>10	trace	trace	1 ml	6

[a]trace separation denotes <0.5 ml

Dry Flowables: The results of evaluations with various commercially available dry flowables is shown in Table 3. There was no definite trend observed in the data. Dry flowables 2 and 3 exhibited unacceptable 24 hour redispersion at all temperatures. For the purpose of this study, >10 inversions was considered unacceptable. In the case of formulations that exhibited acceptable redispersion after 24 hours at 25°C (DF's 1 and 5), there was a definite increase in the number of inversions required to resuspend as the temperature was decreased.

TABLE 3 -- Results for commercial dry flowables

Water	77°F (25°C)				40°F (4.4°C)			
	Observed separation			Inv. to	Observed separation			Inv. to
	15 min	1 hr	24 hr	suspend	15 min.	1 hr	24 hr	suspend
DF-1								
35 ppm	0.5 ml	1 ml	6 ml	10	1 ml	0.5 ml	15 ml	>10
342	0.5 ml	1.5 ml	6 ml	8	1 ml	0.5 ml	15 ml	>10
1000	0.5 ml	2 ml	6 ml	6	1 ml	0.5 ml	15 ml	>10
DF-2								
35 ppm	1 ml	1.5 ml	3 ml	>10	1 ml	1.5 ml	3 ml	>10
342	1 ml	1.5 ml	3 ml	>10	1 ml	1.5 ml	3 ml	>10
1000	1 ml	1.5 ml	3 ml	>10	1.5 ml	1.5 ml	3 ml	>10
DF-3								
35 ppm	1 ml	1.5 ml	4 ml	10	1.5 ml	2 ml	4 ml	>10
342	1 ml	1.5 ml	4 ml	>10	2 ml	3 ml	5 ml	>10
1000	1 ml	2 ml	5 ml	>10	2 ml	3.5 ml	6 ml	>10
DF-4								
35 ppm	5 ml	5 ml	7 ml	6	6 ml	6 ml	6 ml	11
342	4 ml	5 ml	7 ml	6	6 ml	6 ml	6 ml	8
1000	7 ml	10 ml	9 ml	8	6 ml	6 ml	6 ml	6
DF-5								
35 ppm	5 ml	10 ml	10 ml	10	6 ml	12 ml	9 ml	>10
342	3 ml	5 ml	15 ml	8	4 ml	6 ml	11 ml	>10
1000	1 ml	25 ml	12 ml	6	2 ml	13 ml	9 ml	>10

The data from the experiment conducted with DF 4 was inconclusive since complete dispersion of the formulation was not achieved within 10 inversions even at 25°C. To maintain consistency in the study, no attempts were made to completely disperse the material prior to determining the stability.

Conclusions

Based on the results of this study it appears that the dilute performance characteristics of EC formulations are the most susceptible to changes in the carrier temperature, followed in order by dry flowables and liquid flowables. There may be some variation within a formulation class depending upon the actual commercial sample being evaluated, but the samples tested in this study indicate that the performance of EC formulations is highly dependent upon the temperature of the water as well as the hardness. In real world situations carrier temperatures will generally be less than the 25°C test temperature. In order to obtain optimum performance at temperatures less than 25°C evaluation of emulsion performance at these lower temperatures and subsequent formulation modifications, if required, would appear to be advisable.

EMULSIFIER ADJUSTMENTS TO EC'S

Experimental Details

The objective of this experiment was to develop several model EC systems without active ingredient to determine what emulsifier adjustments would be required to achieve emulsion performance at near well water temperatures that is similar to the performance observed at 25°C.

Two solvent systems were chosen for investigation, monochlorobenzene (MCB) and a C9-C15 aromatic hydrocarbon. MCB was selected because it is a widely used solvent in pesticide formulation. Due to the increasing enviromental pressure on low flashpoint solvents such as MCB, higher flashpoint solvents are being evaluated as potential replacements. The C9-C15 aromatic hydrocarbon was chosen because it represents one of these higher flashpoint alternatives.

The emulsifier employed for these experiments was a blend of a 60% calcium dodecylbenzenesulfonate (CaDDBSA), a 36 mole castor oil ethoxylate (36 CO), and a 30 mole nonylphenol ethoxylate (30N). These emulsifier components were selected as they are commonly used in emulsifier blends and were readily available in the lab.

Formulations containing solvent and emulsifier only were blended such that good to perfect blooms and less than 0.5 ml separation were observed at 1 hour in all test waters at 25°C with a 5.0% emulsifier loading. These formulations were then evaluated for the same parameters in 7.7°C°F water samples. If necessary, adjustments to the emulsifier blend were made in order to achieve similar performance as was observed with the same solvent system in the warmer waters. For comparison purposes, HLB values were calculated for the emusifier blends that were actually used in each solvent system at each temperature.

Experimental Results

MCB system: For the system containing MCB it was determined that a ratio of the three emusifiers as outlined in Table 4 was required to achieve the performance criteria stated above. This "formulation" (MCB-A) was then evaluated in 7.7°C water and an observable shift of the emulsion balance toward the hard water was observed. The emulsifier balance was re-adjusted using the same three components to achieve similar performance as had been observed in the 25°C test. The adjusted "formulation" (MCB-B) was then evaluated in 25°C water. The results of this study are located in Table 5.

TABLE 4 -- Emulsifier blends for model systems

Emulsifier	WT % in blend		
	MCB-A	MCB-B	C9-C15
60% CaDDBSA	40.0	49.0	52.0
36 CO	29.0	20.0	42.0
30N	31.0	31.0	6.0
HLB (blend)	12.5	12.2	11.0

C9-C15 aromatic hydrocarbon system: For the system containing the C9-C15 aromatic hydrocarbon it was determined that a ratio of the three emulsifiers as outlined in Table 4 was required to achieve the performance criteria outlined above.   This "formulation" was then evaluated in 7.7°C water and no significant shift was observed vs. the results obtained at 25°C and no adjustments were required.   The results from this study are located in Table 5.   The same criteria were used to evaluate blooms and stabilities as were used for Table 1.

TABLE 5 -- Emulsion results for model EC systems

Water	77°F (25°C)				45°F (7.7°C)			
	Bloom	Emulsion separation			Bloom	Emulsion separation		
		15 min	30 min.	1 hour		15 min.	30 min.	1 hour
MCB-A								
35 ppm	perfect	trace	trace	0.5 ml	poor	1.5 ml	3.0 ml	5.0 ml
342	perfect	0	trace	trace	poor	1.5 ml	3.0 ml	5.5 ml
1000	perfect	0	0	0	good	1.0 ml	2.0 ml	4.0 ml
MCB-B								
35 ppm	poor	0.5 ml	1.0 ml	2.0 ml	perfect	trace	0.5 ml	0.5 ml
342	poor	1.0 ml	1.5 ml	2.5 ml	perfect	trace	trace	0.5 ml
1000	poor	1.5 ml	2.0 ml	3.0 ml	perfect	0	0	trace
C9-C-15								
35 ppm	floats[a]	0	trace	trace	good	trace	trace	trace
342	floats	0	0	trace	perfect	trace	trace	trace
1000	floats	0	0	0	perfect	trace	trace	trace

[a]less dense than water at 25°C

HLB calculation: The HLB value for the emulsifier blends outlined in Table 4 were calculated and are included in Table 4.   The HLB values for the emulsifier blends were calculated as the sum of the individual HLB values of the components multiplied by their percentage in the blend. For the purposes of these calculations HLB values of 9.0 for CaDDBSA, 12.5 for 36CO and 17.0 for 30N were employed.

Conclusions

The results obtained in this study indicate that a higher HLB value is required for MCB based solvent systems as compared to systems based on aromatic hydrocarbons.   The data also indicates that a lower HLB is required to obtain similar emulsion performance when the temperature is decreased from 25°C to 7.7°C.

The data also indicates that a solvent system based on MCB is more sensitive to changes in carrier temperature than one based on a C9-C15 aromatic hydrocarbon system.   Two possible explanations are presented. The first is the fact that MCB is a single component solvent (other than minor impurities) while the aromatic hydrocarbon is a mixture of different molecular weight molecules.   This mixture of molecules may allow the hydrocarbon system to be more resistant to changes in water temperature than the single component MCB system.

A second possible explanation may be the difference in the solubilities of the emusifier blends used in the two systems. The blend used in the aromatic hydrocarbon system contained a higher level of less water soluble components while the MCB blend contained a higher level of more water soluble components. The solubility of the MCB blend would therefore be expected to vary significantly depending upon the water temperature, thus resulting in different emulsion results as the temperature of the water is changed. No effort was made to determine the actual water solubilities of any of the emulsifier blends.

Note that this testing was carried out without active ingredient and that different technicals can be expected to have varying impacts on the emulsion performance.

GENERAL COMMENTS AND CONCLUSIONS

During the course of this investigation no significant difference was observed in the performance of either liquid flowable or dry flowable formulations as a function of water temperature. Significant differences were observed, however, for all commercial EC formulations evaluated, as well as for the model MCB based solvent system. The model aromatic based system showed little difference in performance as a function of water temperature, but the emulsion performance of commercial aromatic based EC formulations has been observed to shift with changes in temperature both in this study and in other lab evaluations.

The observed temperature effect in the case of EC formulations may or may not be important depending upon the user's interpretation of the results. If this test and the results obtained are only being used as a quality control tool to detect and correct for variations in production batches then the relationship between temperature and performance may not be that critical to the effective use of the test. If, however, the purpose of the test is to predict actual end-use performance, then testing at temperatures closer to those actually observed in the field would appear to be advisable.

REFERENCES

[1] "Standard Test Method for Emulsification Characteristics of Pesticide Emulsifiable Concentrates," 1989 Annual Book of ASTM Standards , Vol. 11.04, American Society of Testing and Materials, Philadelphia, 1989.

Nagui I. Ibrahim, Dev K Mehra

COMPRESSED TABLETS AS A POTENTIAL PESTICIDE DELIVERY
SYSTEM

REFERENCE: Ibrahim, N. I. and Mehra, D. K., "Compressed Tablets as a Potential Pesticide Delivery System," Pesticide Formulations and Application Systems: 11th Volume, ASTM STP 1112, Loren E. Bode and David G. Chasin, Eds., American Society for Testing and Materials, Philadelphia, 1992.

ABSTRACT: Solid and liquid pesticides can be formulated into tablets as a delivery system. Inclusion of Lattice™ microcrystalline cellulose (MCC), a special grade of MCC for agricultural application, and Ac-Di-Sol[R] croscarmellose, a cross-linked sodium carboxy methylcellulose (XL-CMC), into the formulation provides a rapid tablet disintegration time of not more than 5 minutes. Lattice MCC acts as an excellent dry binder when the pesticide is compressed thus imparting desired hardness to the tablets. Simultaneously, the XL-CMC being an excellent disintegrating agent, provides the desirable rapid tablet disintegration time. The tablets disintegrate to prime particles and do not form a hard pack when the suspension is left undisturbed for 1-3 days.

KEYWORDS: excipients, disintegration, compression, tablets, hardness

Safe handling of pesticides and the correct application rate are two issues of constant concern to the formulator. To date no true unit dose pesticide formulation that offers flexibility is available on the market. Compressed tablets that can be subdivided to deliver fractional amount of pesticide can be formulated in a bisected or multi sected tablet (Slide 1). Such product form is suitable for the home and garden, green house, forestry, pest control operation and farm markets. In addition, chemically incompatible pesticides can be separated by formulating multilayered tablets (Slide 2). Actives commanding a premium are a good example of products which can be formulated into tablets.

Mr. Ibrahim is Associate Research Chemist and Dr. Mehra is Manager, Pharmaceutical Technical Services at FMC Corp., Box 8 Princeton, NJ 08543.

In order for pesticides, with or without diluents, to be compressed into tablets, it is necessary that the ingredients in the formulation possess a few desirable physical characteristics. These include:
   o The ability to flow freely
   o Cohesiveness
   o lubrication (Slide 3)

Since most materials have none or only some of these properties, methods of tablet formulation and preparation have been developed to impart these desirable character- istics to the material which is to be compressed into tablets. The choice of excipients depend upon a number of factors including the method by which the tablets are to be manufactured. There are three general methods of tablet preparation (Slide 4) [1]:
   1. Wet granulation
   2. Dry granulation
   3. Direct compression

After compression, the tablets must have a number of desirable attributes such as good appearance, requisite hardness, rapid disintegration and technical content un- iformity (Slide 5).

Besides the pesticide or other active agent, other ingredients contained in a tablet formulation are:
   1. Binder
   2. Filler
   3. Lubricant
   4. Disintegrant, etc. (Slide 6) [2].

Binders in a tablet are agents used to impart cohesive qualities to the powder blend, ensuring the tablet remains intact after compression. Binders also improve the free flowing quality of the formulation. The quantity of binder used has considerable influence on the characteristics of the compressed tablets.

BINDERS can be classified as follows: (Slide 7 & 8)

o SOLUBLE

Hydrous Lactose	Sucrose
U.S.P. Crystalline	Dipac^R
or Powder	Nutab^R
Spray Dried	Destab^{TM}
Fast Flo^{TM}	
	Maltodextrin
Anhydrous Lactose	Maltrin^R
DT (Beta)	
DC Lactose 21 (Beta)	Mannitol
DC Lactose 30 (Alpha)	
Tablettose^R	Sorbitol
Zeparox^R	
	Fructose
Dextrose	TAb-Fine F94-M
Emdex^R	
Sweetrex^R (30% Fructose)	Xylitol
Tab-Fine D97-HS	

o INSOLUBLE

ORGANIC

Starch
  Corn
  Potato
  Rice
  Wheat

Starch (Partially
  Pregelatinized)
PH102
  Starch 1500[R]

Whey
  Krafen[R]

Celluloses
  Floc
  Solka[R] Floc
  Elcema[R]
  Keycel[R]

Microcrystalline
  Avicel[R] PH101 &

  Emcocel[R]
  Ex-cel

INORGANIC

Dibasic Calcium Phosphate
  Emcompress[R]
  Di-tab[R]

Tribasic Calcium Phosphate
  Tri-Tab[R]

Calcium Sulfate (Terra Alba)
  Delaflo

Co-Processed
  Celocal[R]
  Vitacel[R]

LUBRICANTS/GLIDANTS have a number of functions in tablet manufacture. Lubricants improve the flow properties of the powder or granules, prevent adhesion of the tablet material to the die and punch surfaces, reduce interparticle friction and facilitate the ejection of the tablets from the die cavity. The effect of glidants is limited to improving the flow properties of powder or granules. The following is a list of lubricants that are used (Slide 9):

LUBRICANTS

	Conc. (%)	Water Sol.
Magnesium Sterate	.25 - 1.5	I
Calcium Sterate	.25 - 1.5	I
Stearic Acid	1 - 4	I
Hydrogenated Vegetable Oils		I
Sterotex[R]	2 - 5	
Lubritab[R]		
Cutina[R] HR		
Mineral Oil	1 - 3	I
Glyceryl Palmitosterate	2 - 5	I
Precirol		
Glyceryl Behenate	2 - 5	I
Coimpritol		
Polyethylene Glycol 6000	2 - 5	S
Sodium Benzoate	1 - 4	S
Sodium Lauryl Sulfate	2 - 3	S
dl-Leucine	1 - 3	S
Talc (antiadherant)		

GLIDANTS

Fumed Silicon Dioxide	.1 - .25	I
Silica Hydrogel	.25 - .5	I

The level of lubricant in the formulation is crucial. Since most lubricants are hydrophobic, they tend to interfere with particle/particle bonding, thus decreasing the compressibility of the powder blend. In addition, the entry of water into the tablets is retarded and thus the disintegration time of the tablet is prolonged. Very rarely do formulations contain more than one percent lubricant.

Mixing time of the lubricant with the ingredients in the tablet formulation is critical. Excessive blending time of more than five minutes will cause the lubricant to be distributed widely over the surface of other ingredi-ents in the formulation and will thus interfere with compression (Slide 10, Figure 1).

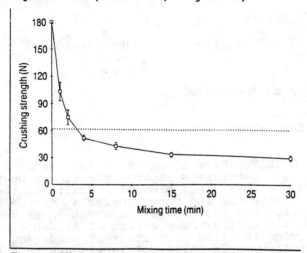

Figure 1 : *Effect of lubricant mixing time on the crushing strength of tablets compressed from the test formulation mixed in a 1000-L V-shaped mixer at 22 rpm.*

Immediately after compression, most tablets will exhibit post compression relaxation and will expand radially and axially within the die cavity and will therefore bind and stick to the sides of the die wall. The choice of a proper lubricant will effectively overcome this friction and will facilitate tablet ejection from within the die cavity.

DISINTEGRANTS are added to the tablets to facilitate breakup. Factors that can influence the disintegration time of a tablet are: Binder level, tablet hardness, lubricant level, type and level of disintegrant, and melting point of active. Typical disintegrating agents used are (Slide 11):

TRADITIONAL   NEW
 Starch (Corn & Potato)   Sodium Starch
Glycolate
  Amylose   Primojel

Alginic Acid	Explotab[R]
Microcrystalline Cellulose	Ac-Di-Sol[R]
Ion Exchange Resins	croscarmellose
Amberlite[R]	Crospovidone
Effervescent Systems (CO_2)	Polyplasdone XL[R]
	Kollidon[R] CL
	Soy Polysaccharides
	Emcosoy[R]

The evolution of carbon dioxide in effervescent tablets is also an effective way to cause the disintegration of compressed tablets. The major drawback is that such formulations must be processed, manufactured, packaged and stored in strictly controlled environmental conditions of temperature and humidity (Slide 12 & 13, Figure 2 & 3) [3].

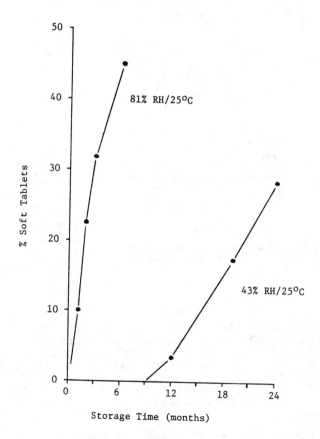

FIGURE 2

Effect of High and Moderate Humidity at Room Temperature
on the Number of Tablets Softening During Storage.

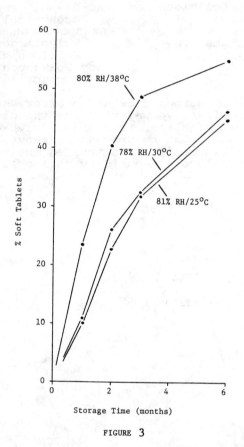

FIGURE 3

Tablet Softening During Six Months of High Humidity
Storage at Various Temperatures.

Direct compression (Slide 14, a review of Slide 4) is the
most desired form of tablet compression provided the
formulation blend demonstrates good flow characteristics
and density. Tablets should be reasonably hard with no
lamination or capping and consistent in weight (Slide
14). This type of compression may not be suitable for
most of the pesticide formulations due to poor mass flow
of the powder.

Dry granulation or slugging involves the compression
of the powder blend to form slugs if a tablet press is
used or compacted sheets if a roller compactor is used.
The resulting slugs or sheets are passed through a 12, 14
or 16 mesh sieve. Final tableting is performed on these
densified granules. This method is suitable for Agricultural products and is also the method of choice when the
active is sensitive to the granulation process.

Wet granulation as a tableting process is not too different relative to the methodology of water dispersible granules. The dried granules are blended with a disintegrant, and lubricant and then subjected to tableting.

It is important to note that the ability of a material to flow uniformly and reproducibly into a die cavity is the primary factor in achieving tablets of equivalent weights.

Selecting the Proper Excipient.

Pesticide tablets require the active, surfactants and part of the diluents to be co-micropulverized to particle size not greater than 30 microns (μm) prior to tableting. The effectiveness of a tablet binder is determined by its dilution potential or carrying capacity which is defined as the weight of ingredients e.g., pesticides, surfactants, etc., which can be carried per unit weight of the binder and can be compressed into tablets of desired hardness and crushing strength (Slide 15, Figure 4).

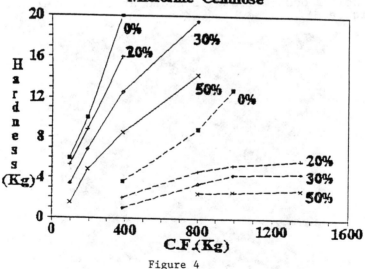

Effect of Ascorbic Acid Concentration on the Compression of Microcrystalline Cellulose and Microfine Cellulose

Figure 4

_____ Microcrystalline Cellulose ---- Microfine Cellulose

The fact that pesticides have to be micropulverized or air milled, results in a tremendous surface area to be compressed. Therefore, the most efficient tablet binder is desirable here, in order to maximize the percentage of active into the tablet. FMC's Lattice™ MCC, designed specifically for the Agricultural industry, meets such requirements and at the same time provides rapid dispersibility in water (Slide 16, Figure 5).

Figure 5:   Scanning Electron Micrograph of Lattice™

Use of a water soluble binder alone will result in tablets having slow disintegration time because the disintegration takes place by erosion only. For the tablets to disintegrate rapidly, they must wick in water. This type and rate of water wicking is demonstrated by Lattice™ MCC. To further expedite tablet disintegration, Ac-Di-Sol[R] croscarmellose is added at 1-2% level (Slide 17, Figure 6).

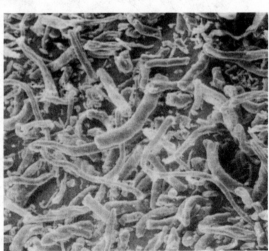

Figure 6:   Scanning Electron Micrograph of Ac-Di-Sol[R]

Pharmaceutical studies have demonstrated that the disintegrant property of Ac-Di-Sol[R] is not affected by storage at elevated temperatures when incorporated into tablet formulations. In Ac-Di-Sol[R] 70% or more of the carboxyl groups are in the sodium salt form ($^-CH_2$-COONA). This confers good hydrophilicity upon the material but because the structure is internally crosslinked, Ac-Di-Sol[R] is insoluble in water. Upon tableting, the internal

pores of Ac-Di-Sol[R] are compressed. When the tablet is
placed in water, the water penetrates into the Ac-Di-Sol[R]
cavities/pores causing an instantaneous expansion thus
resulting in the tablet to break-up rapidly. Ac-Di-Sol[R]
carries a slight negative charge.

Factors to consider in selecting a disintegrant
(Slide 18).

1. Low level use
2. Swelling in water without gelling
3. Surface area
4. Surface morphology
5. Effect of lubricants, glidants disintegrants
6. Flowability alone and in formulation
7. Particle size distribution
8. Moisture content and type and hygroscopicity
9. Bulk density
10. Compatibility with active ingredient
11. Solubility in water

Tablets can vary in size and shape from .1587 cm to
7.62 cm or larger in diameter. Punch faces vary from
flat face to bevelled edge, to standard concave to deep
concave, etc.

Tablet machines can be single punch or rotary presses. In single punch machines the upper punch is mobile.
In the rotary press the upper and lower punches are moving simultaneously during compression. Tablet presses
available are:

1. Kikusui[R]
2. Hata
3. Stokes[R]
4. Killian
5. Courtoy
6. Manesty[R]
7. Fette[R]

Some machines are equipped with precompression and
forced feeding systems. Ideally, the press sould be in-
strumented to measure forces related to the compression,
ejection and detachment of tablets.

Physical properties of the tablets which are rou-
tinely monitored include:

1. Thickness
2. Hardness* (Slide 19, Figure 7)
3. Wt. variation
4. Disintegration time (only test which is com-
     pendial)
5. Friability (Slide 20, Figure 8)
6. Suspension test
7. Residue on 100 mesh screen

* Tablet hardness is a measure of the ability of a tablet
   to maintain its integrity once the compression force
   has been removed. This value is highly dependent upon
   the geometry and size of the tablet. To normalize for

differences in tablet dimensions, tensile strength is a better value to compare among formulations.

$$T.S. = \frac{2H}{\pi DT}$$

where:

T.S. = Tensile strength Kg/cm^2
H   = Hardness in kg
D   = Tablet diameter in cm
T   = Tablet thickness in cm

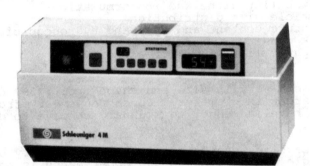

Figure 7 Tablet Hardness Tester

Figure 8 Friability Tester

EXPERIMENTAL

Use of Lattice NT in Agricultural Formulations.

Atrazine technical grade from Ciba-Geigy was formulated into 60g tablets containing 50% a.i.; the formulation was (Slide 22):

		%
1.	Atrazine (95%)	53.15
2.	Lattice™ NT*	36.70
3.	AC-Di-Sol^R	1.4
4.	Dispersant	3.0
5.	Wetting agent	0.6
6.	Synthetic Silica	2.85
7.	Magnesium stearate	0.3
8.	Talc	2.0
		──────
		100.0

* Lattice™ NT is an FMC/FPPD product designed for use as a binder in tableting agricultural products. The binder is 3X more compressible than lactose [Figure 9]. This is an essential characteristic if high AI loading capacity is desired in the tablet formulation e.g., 50% or higher. In addition, Lattice™ NT tends to loose some compressibility if magnesium stearate is used as the tablet lubricant, but remains 2X more compressible than anhydrous lactose [Figure 10]. The physical properties of Lattice™ is illustrated in Table 1.

Procedure:

1. Ingredients 1, 3, 4, 5 and 6 were mixed in a PK^R blender for 10 minutes.

2. The blend was milled in a Fitzmill^R fitted with 0.027 screen openings (knives).

3. Lattice MCC was added and mixed for 5 minutes.

4. The product was slugged on a colton 250 rotary press fed with a vibra screw feeder.

5. Slug dimension was 3.01625 cm in diameter, 0.635 cm in thickness, and 4 g in weight. The slugs were compressed to 6-8 kg hardness as tested with the morehouse hardness tester.

6. The slugs were crushed through 10 mesh screen using an oscillating granulator.

7. The granules were mixed with magnesium stearate and talc for 5 minutes in a PK blender.

8. The mixture was then compressed on the Stokes^R R press single punch machine using 6.6675 cm diameter tooling with tapered dies, at the rate of 16 tablets/min.

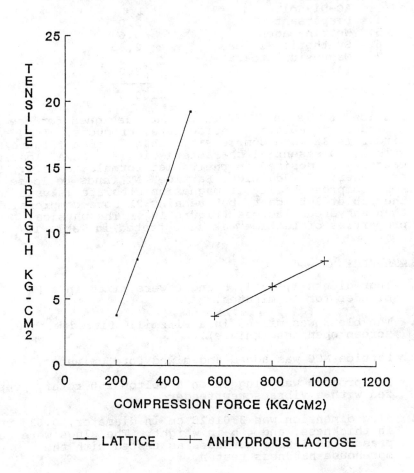

FIGURE 9: COMPRESSIBILITY OF UNLUBRICATED
LATTICE & ANHYDROUS LACTOSE

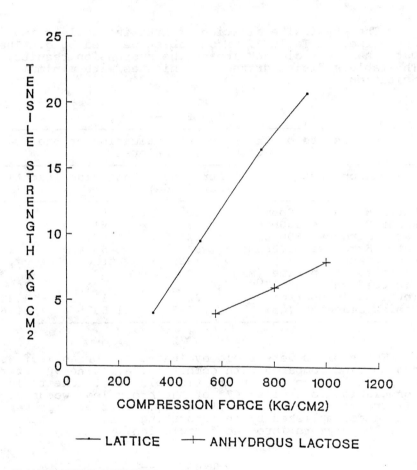

FIGURE 10: COMPRESSIBILITY OF LUBRICATED
LATTICE & ANHYDROUS LACTOSE

1% MAGNESIUM STEARATE

TABLE 1 --

Physical Properties of Lattice™	
Mass Flow (kg/min)	1.66
Volumetric Flow (L/min)	3.67
Loose Density (g/cc)	.45
Tapped Density (g/cc)	.63
% Loss on Drying	1.72
X Particle Size (μm)	110

The physical evaluation of Atrazine tablets is summarized in Table 2. Each tablet weighed 60 g. The water hardness did not impact the suspension results. The tablets disintegrated in 2 minutes with minimum agitation.

TABLE 2 --

Results of the Physical Evaluation of the Compressed Tablets

Suspension Test	Atrazine (Herbicide)	Fungicide	Insecticide
Water Hardness 50ppm	55	51	...
Water Hardness 300ppm	52	51	...
Water Hardness 600ppm	52	50	...
Water Hardness 1000ppm	52	50	...
+ 100 mesh	trace	trace	...
Disintegration time (min) in cold water at 17°C	2	2	3
Total Hardness (kg)	22	11	22
Tablet Diameter (cm)	6.6675	1.5875	6.6675

The tablets were field evaluated for broad leaf weed control and compared with commercial atrazine 90 water dispersible granules or dry flowable (DF). The tablet formulation controlled 67% of the broad leaf weeds at 1lb AI/acre application rate [Table 3]. Further improvement can be accomplished by incorporating nonionic water soluble polymers into the formulation.

TABLE 3 --

Broad Leaf Weed Control 32 Days After Applications

	Lbs Herbicide AI Per Acre		
	0.5	1	1.5
Atrazine 50 Compressed Tablets	57	67	81
Atrazine 90 DF	75	95	95

The tablet formulation was less injurious to corn or soy bean than the dry flowable [Table 4, 5].

TABLE 4 --

Percent Stunting Corn Field Injury
32 Days After Applications

| | Lbs Herbicide AI Per Acre | | |
	0.5	1	1.5
Atrazine 50 Compressed Tablets	0	0	1
Atrazine 90 DF	1	0	0

TABLE 5--

Percent Stunting Soy Bean Injury
32 Days After Applications

| | Lbs Herbicide AI Per Acre | | |
	0.5	1	1.5
Atrazine 50 Compressed Tablets	1	3	5
Atrazine 90 DF	6	13	16

CONCLUSION

Pesticides can be formulated into rapid disintegrating tablets if the proper binder and disintegrant are used. FMC's Lattice™ microcrystalline cellulose product is particularly suited for such application. The product is easy to handle and is EPA approved for Agricultural application under section 180-1001.

ACKNOWLEDGEMENT

The authors acknowledge the efforts of Meena Chaudhary, Joe Middione and Lynn DiMemmo for their contribution in generating data presented in this paper.

REFERENCES

[1] King, R.E., "Remington's Pharmaceutical Sciences", 13th edition, 1965, pp 562-575, Philadelphia, PA.

[2] Shangraw, R.F., "Introduction to Tablets and Tablet Technology", University of Maryland, School of Pharmacy, 1989.

[3] David, S.T. and Gallien, C.E., "Drug Development and Industrial Pharmacy", Vol. 12, No. 14, 1986, pp 2541-2550, Evansville, IN.

David Z. Becher

PESTICIDE COMPATIBILITY: THE EFFECT OF CARRIER

REFERENCE: Becher, D. Z., "Pesticide Compatibility: The Effect of Carrier," Pesticide Formulations and Application Systems: 11th Volume, ASTM STP 1112, Loren E. Bode and David G. Chasin, Eds., American Society for Testing and Materials, Philadelphia, 1992.

ABSTRACT: The routine use of pesticide tank mixtures has increased the importance of formulation tank mix compatibility. Unfortunately, compatibility is very sensitive to the mixing and spraying conditions. Consequently, incompatibility occurs erratically in the field and is difficult to consistently reproduce in the laboratory. While compatibility in the field can never be totally predicted in the laboratory, the test method used in this work provides good predictions, while still allowing the testing of many combinations of pesticides and carriers with a reasonable effort.

   It is known that the results of compatibility tests are dependent on the spray carriers used. It was not known whether tests with laboratory carriers would adequately predict field results. Therefore, samples of water and ammonium nitrate/urea liquid fertilizers were obtained from various parts of the United States and their effect on the compatibility of pesticide tank mixtures evaluated. Samples of nominally similar carriers gave very different compatibility results. The importance of using carriers which are representative of the ones used in the field when evaluating pesticide tank mix compatibility was confirmed.

KEY WORDS: compatibility, test method, tank mix, pesticide, formulation, liquid fertilizer, water

David Becher is a Senior Research Specialist at Monsanto Agricultural Company, A Unit of Monsanto Company, 800 N Lindbergh Blvd. Saint Louis, Mo. 63167

INTRODUCTION

The importance of the tank mix compatibility of agricultural formulations has been recognized for twenty years [1]. Unfortunately, the accurate reproduction of field compatibility problems is not easy. A good laboratory compatibility test method is, however, very valuable. For a new formulation, a good compatibility method can make unnecessary the large scale testing of an unacceptable product. When compatibility problems are discovered with an existing product, a good test method is of value in determining both the cause of and possible cures for the problem.

The spraying of pesticides as tank mixtures is common, since a single pesticide, generally, will not control all the pest species present in a particular field. For example, an applicator may apply two pre-emergent herbicides, one to control grasses and a second for broad-leaved weeds. In addition, he might add other pesticides, to kill weeds that have already germinated, for example. Furthermore, synergistic effects are sometimes observed when two pesticides or a pesticide and an adjuvant are applied together. It is clearly more convenient and efficient for an applicator who is applying more than one product to spray them simultaneously.

There are three basic types of incompatibility: biological, chemical, and physical [1]. Biological incompatibility occurs when a mixture of products has less activity when they are applied together (antagonism) or causes injury to desirable species that are not harmed when the products are applied separately. Chemical incompatibility occurs when an undesired chemical reaction, such as the formation of an insoluble or biologically inactive salt, occurs between two pesticides and reduces their activity. Careful reading of the products' labels should prevent these problems.

The most difficult type of compatibility problem to deal with is physical incompatibility. It is observed when the inert ingredients of a formulation interact in an undesirable manner with another formulation or the spray carrier. This type of incompatibility tends to be sporadic in the field, since its occurrence is often dependent on the use conditions. The most commonly observed problem is the formation of solid or semi-solid lumps or curds, which can plug spraying equipment. Both the equipment used and how the product is handled can have a significant effect on the probability and severity of the incompatibility [2].

The reproduction of compatibility problems in the laboratory is similarly very sensitive to the test conditions and materials used. Brenner et al. [3] examined the effect of water hardness, temperature, copesticide ratio, addition order, agitation, residence time, and dilution rate on the results of a laboratory compatibility test. The water hardness was, by far, the most significant variable. However, major effects were also observed for several other variables including agitation, residence time, and dilution rate. Also, some of the interactions of these variables were significant. The authors found that both the relative importance of the variables and their effects were dependent on what formulations were used in the tests. They concluded that a good predictive laboratory test was impractical because the conditions in the field vary so widely that the number of combinations that would have to be tested exceeds the usefulness of the results.

It is certainly impossible to test or even simulate all the conditions that a formulation may experience in the field. In consequence, it is possible that testing in the laboratory will fail to detect a significant compatibility problem. However, good compatibility is a vital property for a formulation and extensive testing is justified. A test that can predict the performance of a formulation under the most commonly encountered conditions can be of value when used as a part of a well designed test program.

Devisetty et al. [4] investigated the effect of different grades of fertilizer on the compatibility of a few pesticide mixtures. They tested three types of fertilizer. They found that most combinations were incompatible, unless an adjuvant was used. Mixtures that were unsprayable and those that merely separated on standing without agitation were not clearly distinguished. Also, the effects of variations within a single type of fertilizer were not investigated.

In previous investigations such as Devisetty et. al., the experiments were usually carried out with samples of water and liquid fertilizer from a limited number of sources. Usually the carriers tested were formulated in the laboratory rather than being commercial products. The carriers used in the field vary far more than the standard waters and model fertilizers. Wide variations in composition and properties can be found between products that are nominally the same. In this work, the effect of these variations on the compatibility of pesticide tank mixtures was examined.

CARRIERS

In the real world, the variety of liquids from which pesticides are sprayed is very large. Both water of a wide range of hardnesses and a large variety of liquid fertilizers are used. In the United States, the natural water hardness varies from 3000 ppm or more in some areas of West Texas and the Dakota Badlands to less than 5 ppm of $CaCO_3$ for some wells in the Mississippi Delta [5].

Besides water, the use of liquid fertilizer solutions as pesticide carriers is now routine. The common commercial liquid fertilizers can be divided into six types:

1. Anhydrous ammonia
2. Aqueous ammonia
3. Ammonium nitrate/urea (URAN)
4. Ammonium nitrate/urea/ammonia
5. Ammonium nitrate/urea/sulfur (SURAN)
6. Liquid mixed (NPK) fertilizers

Of these, the ammonium nitrate/urea solutions are the ones in which pesticides are most often applied. However, the use of sulfur containing fertilizers is increasing. In contrast, the application of pesticides in anhydrous or aqueous ammonia is rare.

The compositions of the ammonium nitrate/urea fertilizers are usually identified by the weight per cent of nitrogen in the solution. There are many different possible mixtures, but the 28% and 32% nitrogen solutions are the most commonly used. The exact composition of these fertilizers depends on the manufacturer, but the usual practice is to use a ratio of the ingredients that will minimize the salting-out temperature. A typical 28% nitrogen fertilizer, therefore, contains about 40% ammonium nitrate, 30% urea, and 30% water and has a salting out temperature of -18ºC (-1ºF). The fertilizers containing sulfur are even more varied as the sulfur can be added in large variety of forms. A typical example is a 28-0-0-4S fertilizer, which contains 28% nitrogen and 4% sulfur.

COMPATIBILITY TESTING

The most basic and simplest compatibility test method is the so-called jar test [2,6]. It was originally developed to determine the effectiveness of tank-mix compatibility agents. The basic test consists of filling two one-quart jars about half full of the carrier liquid to be used, and adding the pesticides to be sprayed in the same order and at the same concentration as they will be added to the actual spray solution. The compatibility agent to be tested is then added to one jar while the other is used as a control. The jars are shaken to mix the ingredients and allowed to stand. The resulting mixtures are examined after 5 and 30 minutes. The relative compatibility of the two mixtures can be compared to determine if the adjuvant is effective.

This test is simple and requires no equipment. Applicators have used it for many years, since it can be run at the application site to check compatibility immediately before application. If any problems are detected, compatibility agents can be added or other corrective actions taken.

This method is useful for evaluating the large effects of the addition of a compatibility agent. However, it is not a particularly reliable method of predicting if a pesticide mixture will give adequate compatibility in the spray tank. The test conditions are too poorly controlled and do not adequately simulate the actual conditions under which the pesticides are used. Additionally, when used to evaluate tank mix compatibility, the method is being used to make a measurement of absolute instead of relative compatibility. In consequence, this method frequently does not predict or duplicate problems.

At the opposite end of the spectrum in complexity, are test methods that try to simulate an actual spray rig. An example is the Stepan Laboratory Sprayer [7], which is essentially a miniature spray rig. With this type of equipment, it is theoretically possible to reproduce field conditions, since it should be possible to duplicate the flow rates and agitation of a real spray rig. However, this method also has significant drawbacks. It requires large amounts of material; it is time-consuming; and it can be difficult to clean the equipment. The testing of more than a few combinations with this type of equipment is usually not practical.

Fortunately, a test method that is more reproducible than the jar test and easier to perform than the simulated spray rig is possible. It is a refinement of the jar test with better control of the conditions and a somewhat more quantitative evaluation of the results. This test is an attempt to address several shortcomings of the jar test including:

1. The mixing is not reproducible.
2. The addition rates of the pesticides are not controlled.
3. The samples tend to separate into layers on standing.
4. Spray tank agitation is poorly simulated.
5. The evaluation of compatibility is not quantitative.

This test was developed in an effort to duplicate field complaints. The equipment and materials needed are:

Burrell Wrist Action Shaker equipped with shaker arms suitable for 2
    ounce (59 mL) bottles (Burrell Corp.)
Bottles, Two ounce (59 mL) Boston Round and caps or equivalent
Graduated cylinder, 50 or 100 milliliter
Pharmaceutical graduate, five or ten milliliter
A balance accurate to at least 0.01 gram
A Vari-pet syringe with Teflon Tip (Fisher)
A three inch diameter 50 mesh stainless steel screen
The carrier solutions to be used
The pesticides to be tested

The procedure for the test is:
•   To each of the required number of two ounce bottles, is added 50 milliliters of one of the spray carriers being used.

•   To each bottle, is then added the first pesticide. The pharmaceutical graduate is used to measure the liquids and the balance is used to weigh the solids. The amounts added are chosen so the ratio of the pesticides is the most typically used value and the total amount of pesticide added is approximately 4 milliliters. The order of pesticide addition is the one recommended on the product's labels.

It is important to attempt to hold the total amount of pesticide used relatively constant as the amount of empty volume in the bottle is a significant variable. The total amount of pesticide used was chosen to simulate the higher application concentrations normally used since this would usually be the most difficult conditions to obtain compatibility.

•   The bottle is then capped and swirled gently until the pesticide is dispersed as well as possible. If it is not possible to test all the samples containing one carrier in a single test, the samples are divided so all the bottles with a particular first pesticide are tested together.

•   The caps are then removed and the second pesticide is added. For liquids the Vari-pet adjustable syringe is used, while solids are weighed onto a piece of paper and poured into the bottle. The results can be sensitive to the method of addition of the second pesticide. Therefore, an effort should be made to add it in a reproducible way. The best technique appears to be to add the second pesticide as quickly as possible.

•   The bottles are then capped and gently swirled until their contents appears to be homogeneous.

•   They are then placed on the Burrell Shaker. The shaker arms are adjusted so the bottles make an angle of between 20 and 45 degrees with the vertical.

•   The bottles are then shaken for 30 minutes at a shaker setting of ten (maximum arm movement).

•   After shaking is completed, the bottles are removed from the shaker.

•   Each sample is swirled lightly to redisperse any settled material and poured through the 50 mesh screen. The amount of material retained on the screen and in the bottle is evaluated.

•   If any material is observed to be retained, the bottle is then washed twice by filling it with water and swirling it briefly, and the water is poured through the screen. The effect of this washing on the retained material is observed.

•   The compatibility is then rated on a five point scale:

    1 --- Perfect Compatibility
    2 --- Screen Clean; Some Specks on the Bottle Wall
    3 --- A Small Deposit on the Screen which Washed Through with Water
    4 --- Moderate to Large Amount on the Screen
    5 --- Material Remains in the Bottle

Sometimes a small amount of material resulting from dry flakes in the flowables is observed on the screens. It should be ignored when rating the samples. If there is any doubt, a blank with only one herbicide should be run to confirm that the flakes are from the flowable.

•   The screen is then carefully cleaned and the solutions and bottles disposed of in a safe manner.

This method is quick and easy to run. It is more reproducible than the jar test and requires less material. Up to 16 samples can be run simultaneously on a single shaker. In addition, it is easy to replicate experiments, which increases the chances of observing borderline problems.

MATERIALS

The water and fertilizers used in this study were obtained by requesting samples of actual spray carriers from different areas of the Midwest. It is believed that they are representative of the materials that will be encountered in the field. No attempt was made to either obtain or avoid carriers that had been implicated in compatibility complaints. Nine liquid fertilizers, three samples of water from the field, and three samples of water of standard hardnesses, which were prepared in the laboratory, were used in the experiments. The fertilizers included one sample containing 32% nitrogen, six containing 28% nitrogen, and two 28-0-0-4S fertilizers. The 32% fertilizer and one of the fertilizers containing sulfur were obtained from Arcadian. The other fertilizers were samples from dealers in the Midwest. The laboratory water samples were standard 1000 ppm all calcium water and 28 and 57 ppm waters made by diluting WHO water with deionized water. The three samples of water from the field were from the supplies of farm product dealers.

The pesticides used in this study were divided into two groups. One set, which was used as the first pesticide in the test method, consisted of commercial samples of DuPont (Shell) Atrazine® 4L, Ciba-Geigy Aatrex® 4L, Ciba-Geigy Aatrex® 9-0, and DuPont (Shell) Bladex® 4L. The second group contained five experimental alachlor emulsifiable concentrates. These samples were intentionally chosen to give a range of tank-mix compatibilities. The quantities of each pesticide used in the tests are given in the following Table 1.

TABLE 1-- Pesticide Amounts

Coherbicide	Amount of Coherbicide	Amount of EC
DuPont Atrazine® 4L	1.4 mL	2.5 mL
Aatrex® 4L	1.4 mL	2.5 mL
Aatrex® 9-0	0.8 g	2.5 mL
Bladex® 4L	1.6 mL	2.4 mL

These amounts were chosen by taking the median use rates from the Lasso® label in use at the time of the experiments. The ratio of these rates was then used to calculate the amounts of each pesticide needed to achieve a total pesticide volume of about 4 milliliters.

RESULTS

The compatibilities of the combinations of spray carrier, emulsifiable concentrate, and flowable or dry flowable were determined. For each combination of emulsifiable concentrate and carrier, the ratings for the four coherbicides were added. This produced a rating scale of from 4 to 20 for each combination of carrier and emulsifiable concentrate. For example, if a particular combination of carrier and emulsifiable concentrate gave a rating of 1 with Aatrex® 4L, 4 with DuPont Atrazine 4L, 2 with Bladex@ 4L, and 5 with Aatrex 9-0 the overall rating would be 12. Therefore, a product with the best rating (1) with all four copesticides would have a score of 4 and one with the worst (5) would have a total score of 20. This yielded a single value for each combination and was useful for the statistical analysis. The results of the experiments are summarized in Table 2.

TABLE 2. -- Experimental Results

CARRIER	EC 1	EC 2	EC 3	EC 4	EC 5	MEAN
32% N Sample 1	5.33 (3)	5.00	4.00 (2)	6.00	9.00	5.87
28% N Sample 1	4.00 (2)		4.00 (1)	8.00	7.00	5.75
28% N Sample 2	5.00 (3)	4.00	4.00 (2)	8.00	8.00	5.80
28% N Sample 3	5.67 (3)	6.00	6.50 (2)	5.00	7.00	6.03
28% N Sample 4	5.33 (3)	7.00	7.50 (2)	5.00	7.00	6.37
28% N Sample 5	7.00 (3)	6.00	8.00 (2)	4.00	8.00	6.60
28% N Sample 6	9.00 (3)		11.00 (2)		10.00	10.00
28-0-0-4S Sample 1	7.00 (3)	10.00	7.50 (2)	8.00	6.00	7.70
28-0-0-4S Sample 2	9.00 (3)	11.00	9.50 (2)	12.00	7.00	9.70
28 ppm Water	4.00 (3)	4.00	4.00 (2)	7.00	4.00	4.60
57 ppm Water	4.00 (3)	4.00	4.00 (2)	7.00	4.00	4.60
Indiana Water	4.00 (3)	4.00	5.50 (2)	4.00	7.00	4.90
Missouri Water	4.00 (3)	4.00	4.00 (2)	7.00	7.00	5.20
Illinois Water	4.00 (1)		10.00 (1)		10.00	8.00
1000 ppm Water	5.00 (3)	4.00	8.50 (2)	7.00	13.00	7.50

For the first and third emulsifiable concentrates, the values are the averages of several measurements. The numbers in parentheses are the number of measurements. The column labelled 'Mean' contains the average of the values for the five emulsifiable concentrates.

A trend towards poorer compatibility ratings is observable from EC 1 to EC 5. This was expected as the formulations were intentionally chosen to differ in compatibility performance. Within the fertilizers and waters, there is also a trend towards poorer ratings as one goes down the table. Obviously, the carrier used in a test can have a noticeable effect on the results. As expected, the fertilizers containing sulfur gave poorer ratings than most of the ammonium nitrate/urea ones. The wide range in performance among the 28% nitrogen fertilizers is, however, striking.

When the results for the different carriers are compared, there appears to be a general trend from best to worst performance which is independent of the emulsifiable concentrate. However, there are clearly differences between the emulsifiable concentrates in the relative compatibility with the carriers. In particular, the results for EC 4 appears to differ from the general pattern. It is notable that it is also the formulation which differed the most in composition from the others. This suggests that for formulations of similar composition, it is probable that an easily determinable trend from best to worst can be established when a series of fertilizers are compared using this test. However, the trend may not hold if formulations with larger variations in composition are included in the test.

STATISTICAL ANALYSIS

Replicates for the first and third emulsifiable concentrates, including data from additional experiments not reported in Table 2, were used to estimate the error in the measurements. The standard deviations for the thirty combinations of carrier and emulsifiable concentrate were pooled to get an estimate of 1.43 (100 df).

The data in Table 2 were analyzed using the SAS [8] statistical package using a general linear models method. The results are presented in Tables 3 and 4. Table 3 contains the analysis of variance and Table 4 the results of tests of the significance of the differences in the ratings of the emulsifiable concentrates and carriers tested.

This analysis required two important assumptions. One is that the results can be treated as a continuous variable. The possible values vary from 4 to 20 and the observed ones from 4 to 13. The second assumption was that the unequal number of measurements made for the different combinations would not significantly effect the results. The primary effect of these assumptions is on the significance tests for the differences between the means for the different factors.

Three class variables, the 'EC', the 'Carrier', and the 'Set', and their two factor interactions were used in the analysis. The 'Set' variable is a blocking factor that was used to allow for possible variations between experimental runs. Although the factor is statistically significant, the conclusions do not seem to be changed by its inclusion in the analysis. Tukey's studentized range test was used to distinguish which factors were significantly different.

TABLE 3 -- Statistical Summary

SOURCE	DF	SUM OF SQUARES	MEAN SQUARE	F VALUE
MODEL	95	565.55	5.953	13.79
ERROR	11	4.75	0.432	
CORRECTED TOTAL	106	570.30		

$PR > F^{a} = 0.0001$

R-SQUARE	0.9917		

SOURCE	DF	TYPE I SS	F VALUE	PR > F
EC	4	60.37	34.95	0.0001
CARRIER	14	254.45	42.09	0.0001
SET	2	4.30	4.98	0.0289
EC*CARRIER	51	174.49	7.92	0.0001
EC*SET	1	0.75	1.74	0.2143
CARRIER*SET	23	71.20	7.17	0.0008

[a] Where PR>F is the probability of a value of F greater than the calculated one occurring by random chance

The fit of the data to the model, as tested by the F and R-Square values in Table 3, appears to be acceptable and the results of the analysis in Table 4 agree with the qualitative conclusions from the examination of the data. For the water samples, the performance of the formulations tested wassignificantly poorer in the 1000 ppm and Illinois water samples than in the other waters. For the fertilizers, the compatibility was poorer in the sulfur containing (SURAN) fertilizers than in most of the URAN type. However, a very wide range of compatibility performance is obviously possible as is shown by the differences between the 28% nitrogen fertilizer samples. Of the 28% nitrogen fertilizer samples, the sixth clearly resulted in poorer compatibility than any of the others. There are also significant differences between the other fertilizers, including the two SURAN products.

TABLE 4. -- Results of Tukey's Studentized Range Test

TUKEY GROUPING	MEAN	N	EC
A	7.60	15	EC 5
B	6.77	13	EC 4
BC	6.33	27	EC 3
CD	5.75	12	EC 2
D	5.42	40	EC 1

TUKEY GROUPING	MEAN	N	CARRIER
A	10.00	3	28% N Sample 6
A	9.50	8	28-0-0-4S Sample 2
B	8.00	3	Illinois Water
BC	7.50	8	28-0-0-4S Sample 1
BCD	7.00	8	1000 ppm Water
BCDE	6.88	8	28% N Sample 5
CDEF	6.25	8	28% N Sample 4
DEFG	6.00	8	28% N Sample 3
EFGH	5.50	8	32% N Sample 1
FGH	5.40	8	28% N Sample 1
FGH	5.38	8	28% N Sample 2
GH	4.75	8	Missouri Water
GH	4.75	8	Indiana Water
H	4.38	8	28 ppm Water
H	4.35	8	57 ppm Water

MEANS WITH THE SAME LETTER(S) ARE NOT SIGNIFICANTLY DIFFERENT.

CONCLUSIONS

The compatibility test method described in this work has significant advantages over the standard jar test and other similar compatibility methods. Because it better controls the test conditions, it tends to be more reproducible than the jar test. It is easier and quicker than methods which attempt to simulate a spray rig and requires smaller samples. This allows the testing of a larger number of combinations.

The importance of testing many combinations of carrier and pesticide was clearly demonstrated by this work. The need to test the compatibility in a wide range of water hardnesses was confirmed. It appears that, if laboratory waters covering a large enough range of hardnesses are tested, a good evaluation of the compatibility can be obtained. The testing of water samples from the areas in which the product will be used is, however, still a good idea. Unlike standard laboratory waters, water from the field can contain a wide variety of ions and other solutes, which could affect the results of the tests.

For fertilizers, it is important to test a wide range of sources. If only one or two of the fertilizer samples had been tested, the compatibility of some of the emulsifiable concentrates might have been badly over- or under-estimated. The compatibility properties of the liquid fertilizers available in the field can vary greatly. They are produced by many local suppliers and their compositions and the purity of the raw materials used varies. Generally, the combinations with the fertilizers from the field seemed to give poorer compatibility than the ones made with laboratory fertilizers.

When formulations with compositions that differ greatly are compared, it is even more important to test a wide range of carriers. It is possible that carriers that give good performance with one formulation will give poor performance with another. Therefore, it is not easy to pick a small number of carriers that will cover the entire range of performance. Fortunately, the compatibility test described in this paper makes it easy to quickly test the compatibility of a formulation with several copesticides in a large number of carriers. This allows a formulation to be sent to field testing with reasonable confidence that no unexpected compatibility problems will be encountered.

ACKNOWLEDGEMENTS

The author gratefully acknowledges the assistance of the Monsanto Product Development Department which provided the fertilizer and water samples.

Lasso® is a trademark of Monsanto Company.
Bladex® is a trademark of E. I. DuPont de Nemours and Company.
Aatrex® us a trademark of Ciba-Geigy Corporation.

REFERENCES

[1] Johnson, H. F. and Kaldon, H. E., "Compatibility of Pesticide Tank Mixtures," in *Herbicides, Fungicides, Formulation Chemistry*, Proc. 2nd Int. IUPAC Congr. Pestic. Chem. (A.S. Tahori, ed.), Gordon and Breach, New York, 1972, Vol. 5, pp. 485-521.

[2] Houghton R. D., "Pesticide Compatibility: An Overview from Technical Services," in *Pesticide Tank Mix Applications: First Conference*, (J. F. Wright, A. D. Lindsay and E. Sawyer, eds.), ASTM, Philadelphia, Pa., 1982, pp. 3-10.

[3] Brenner, W. E., Brown, L. J., and Machado, I. E., "Evaluation of Factors Affecting Tank Mix Compatibility of Pesticide Combinations," in *Pesticide Formulations and Application Systems Fourth Symposium*, (T. M. Kaneko and L. D. Spicer, eds.) ASTM, Philadelphia, Pa., 1985, pp. 24-36.

[4] Devisetty, B. N., Zuccarini, R. L., Hanson J. R., Ernst H. W., and Haug W. H., "Compatibility Agents for Liquid Fertilizer-Pesticide Combinations," in *Pesticide Tank Mix Applications: First Conference*, (J. F. Wright, A. D. Lindsay and E. Sawyer, eds.), ASTM, Philadelphia, Pa., 1982, pp. 11-23.

[5] Lindner, P., "Effect of Water in Agricultural Emulsions," in *Herbicides, Fungicides, Formulation Chemistry*, Proc. 2nd Int. IUPAC Congr. Pestic. Chem. (A.S. Tahori, ed.), Gordon and Breach, New York, 1972, Vol. 5, pp. 453-469.

[6] McGlamery M. D., "Sharpen Up Your Tank Mixing Techniques," *Farm Chemicals,* Vol. 142, No. 3, March 1979, pp. 74-6.

[7] Smith, N. R., "Evaluation of Tank Mix Compatibility Using the Laboratory Sprayer," in *Pesticide Formulations and Application Systems Third Symposium*, (T. M. Kaneko and N. B. Akesson, eds.), ASTM, Philadelphia, Pa., 1983, pp. 52-64.

[8] SAS Institute Inc., *SAS® User's Guide: Statistics, Version 5 Edition.* SAS Institute Inc., Cary, N.C. 1985.

Herbert M. Collins and Lawrence A. Munie

Improvements in Dry Flowable Tank Mix Compatibility

REFERENCE: Collins, H. M. and Munie, L. A., "Improvements in
Dry Flowable Tank Mix Compatibility," Pesticide Formulations
and Application Systems: 11th Volume, ASTM STP 1112, Loren E.
Bode and David G. Chasin, Eds., American Society for Testing
and Materials, Philadelphia, 1992.

ABSTRACT: Tank mix compatibility has frequently been a problem for dry flowable
formulations. Compatibility with either co-pesticides or liquid fertilizer has been
poor in the past resulting in flocculation or sedimentation. Traditional
lignosulfonate and naphthalene condensate surfactants used in dry flowables appear
to compound these compatibility problems. Developments in surfactant technology
have served to improve the tank mix compatibility in dry flowables.

A new class of dry flowable surfactants is discussed and compared to the
traditional surfactants. The compatibilities of many dry flowables are extensively
evaluated. The new surfactants were found to improve tank mix compatibility
without the loss of suspensibility in water.

KEY WORDS: Compatibility, dry flowable, fertilizer compatibility, shaker test,
dispersion, suspensibility.

Tank mix compatibility is a serious problem for many dry flowable
formulations. Compatibility with other pesticides and liquid fertilizers has
frequently been inadequate causing flocculation and sedimentation resulting in
either an inability to spray or nonuniform field application coverage. Traditional
sulfonated naphthalene-formaldehyde condensate surfactants used in dry flowables
appear to compound these compatibility problems. Recent developments in
surfactant technology have provided dispersants that improve tank mix compatibility
when using dry flowables.

These new dispersants are discussed and compared to the traditional
surfactants. Compatibility studies are conducted with several major dry flowables
in tank mix combinations with other pesticides and liquid fertilizers.

Mr. Collins is the Agricultural Formulations Manager at Stepan Company. P.O. Box
687, Winder, GA. 30680; and Mr. Munie is an Agricultural Formulations Chemist
at Stepan Company. P.O. Box 687, Winder, GA. 30680

The new dispersants investigated in this study differ in that they are anionic and nonionic surfactant blends absorbed onto a lignosulfonate substrate. Lignosulfonates enhance the dispersancy, but the dramatic improvements in tank mix compatibility are attributed to the additional surfactants. The dry surfactant blends enable the formulator to manipulate the formulation along the lines of an emulsifiable concentrate. This feature was impossible prior to the introduction of these blends.

Compatibility, whether it be of a simple mixture of pesticide and water or a complex blend of herbicide, fungicide and insecticide broadcast in liquid fertilizer, is a concern. With the economic stress placed on farmers, they will continue to mix products to reduce the number of passes through their fields. This practice places a premium on formulations that are compatible. Not only does the tank mixture have to be sprayable but it must be sprayed evenly through the field.

The compatibilities of LASSO° 4E and DUAL° 8E tank mixed with atrazine, cyanazine and linuron dry flowables were evaluated in a variety of liquid fertilizers. Each of the combinations was evaluated using commercially available pesticides. The dry flowables were also prepared using the new dispersants (Formulation 1 for atrazine) and compared to the commercial pesticides. In addition, an atrazine dry flowable was prepared using the traditional sodium naphthalene-formaldehyde condensate dispersant (Formulation 2). Cyanazine 90DF and Linuron 50DF were prepared using the new dispersants (Formulation 3 and Formulation 4 respectively).

ATRAZINE 90DF
New Dispersant

	WT%
Atrazine (98%)	92.0
STEPWET° DF-60	2.0
STEPSPERSE° DF-100	6.0

Formulation 1

ATRAZINE 90DF
Traditional Dispersant

	WT%
Atrazine (98%)	92.0
Dioctyl Sodium Sulfosuccinate	0.5
Sodium Naphthalene Sulfonate	1.5
Sodium Sulfonated Naphthalene-Formaldehyde Condensate	6.0

Formulation 2

LASSO is a registered trademark of Monsanto Company.
DUAL is a registered trademark of Ciba Geigy Corporation.
STEPSPERSE and STEPWET are registered trademarks of Stepan Company.

CYANAZINE 90DF
New Dispersant

	WT%
Cyanazine (97%)	93.0
STEPWET DF-90	1.0
STEPSPERSE DF-300	2.0
Sodium Sulfonated Naphthalene-Formaldehyde Condensate	4.0

Formulation 3

LINURON 50DF
New Dispersant

	WT%
Linuron (95%)	52.6
STEPWET DF-90	3.0
STEPSPERSE DF-300	8.0
WESSELON° 50S	10.0
MIN-U-GEL° 200	26.4

Formulation 4

In order to eliminate the effect of the granulation processes, all the commercial granules were slurried in an attritor to insure the completed disintegration of the granules. All slurries were evaluated for particle size, dispersion and percent suspensibility [2]. The particle size range for all of the slurries was 5 to 15 microns. The results of the percent suspension and dispersion testing are listed in Tables 1 and 2 respectively. The major differences in test results are related to the surfactants used in the formulations. The commercial slurries had a lower percent suspensibility than their experimental counterparts. All pesticides were evaluated at recommended label application rates. The rates are listed in Table 3.

Table 1 -- Percent Suspensibility of the Dry Flowable Slurries

Product	Percent Suspensibility	Product	Percent Suspensibility
Stepan Atrazine	97	Stepan Cyanazine	99
Generic Atrazine	97	Commercial Cyanazine	80
Commercial Atrazine	87	Stepan Linuron	80
		Commercial Linuron	73

WESSELON is a registered trademark of Degussa Corporation.
MINUGEL is a registered trademark of Floridan Company.

Table 2 -- Mixing Cylinder Test for Dispersion

342 ppm Water	Dispersion, %		Redispersion (# Inversions)		
	30 min	60 min	24 hrs	60 min	24 hrs
Stepan Atrazine	0	0	3.5	0	18
Generic Atrazine	2	2	3	5	>20
Commercial Atrazine	1	1	2	3	>20
Stepan Cyanazine	0	0	4	10	18
Commercial Cyanazine	1.5	2	3	10	15
Stepan Linuron	0	0	23 Floc	0	1
Commercial Linuron	3.5	4	4	1	2

1000 ppm Water	Dispersion, %		Redispersion (# Inversions)		
	30 min	60 min	24 hrs	60 min	24 hrs
Stepan Atrazine	0	0	6	0	18
Generic Atrazine	1	2	3	10	>20
Commercial Atrazine	2	3	5	11	>20
Stepan Cyanazine	0	0	30 Floc	0	1
Commercial Cyanazine	0.5	1	3	3	6
Stepan Linuron	0	0	40 Floc	0	1
Commercial Linuron	3.5	4	4	1	2

Table 3 -- Recommended Broadcast Application Rates Per Acre

CO-PESTICIDE	LASSO 4E	DUAL 8E
Atrazine 90DF	2.5qt/2.2lb	1qt/2.2lb
Cyanazine 90DF	2.5qt/1.5lb	1qt/1.5lb
Linuron 50DF	2.5qt/2lb	1qt/2lb

Three methods were used for the evaluation of compatibility in this study. All three methods are currently employed throughout the industry as standards. The first method, the Agricultural Laboratory Sprayer [1], models actual field conditions of agitation, circulation and spraying. This method can be altered to duplicate any configuration used by the farmer. The second method, the shaker test, was used as an alternate dynamic method. This method with screening of the solutions, indicates any clogging of either the in-line strainer or the spray nozzle filters. The third method employed, the mixing cylinder test, provides a crude estimation of compatibility. This method only indicates possible problems, but can not confirm their existence.

The most indicative compatibility method was the Agricultural Laboratory Sprayer (Illustration 1). It simulated actual field conditions and was the most reliable indicator of tank mix compatibility. A 100 mesh in-line strainer and a 50 mesh screen at the spray nozzle were used throughout these tests. The Laboratory Sprayer was set at 30 psi at the discharge outlet. Each test was run for 1 hour with the recirculation line open. At 4 minute intervals a 100 ml mixing cylinder was filled from the spray nozzle. The spray solution collected in each cylinder was monitored for 1 hour for separation. The amount of pesticide separating in each of the cylinders was used to determine the distribution pattern. Incompatibility occured when the filters clogged or the separation in all aliquots were irregular. The test results obtained with the Agricultural Laboratory Sprayer are listed in Table 5.

Illustration 1 -- Agricultural Laboratory Sprayer

Compatibility Test #1   Laboratory Sprayer

Materials

- Laboratory sprayer
- Carrier of choice
- 100 ml mixing cylinder
- Timer

Procedure

1. Fill the spray tank to two thirds full with 1 gallon of carrier.

2. With the carrier circulating add the pesticides, one at a time, allowing for complete mixing between additions.

Note: If more than one pesticide is to be used, add them separately in the following order.

1. Wettable Powders.
2. Water Dispersible Granules.
3. Flowables.
4. Water Solubles.
5. Emulsifiable Concentrates.

3. With the spray mixture circulating, spray a 100 ml aliquot of the mixture every 4 minutes throughout a 1 hour spray cycle.

4. Observe the mixing cylinder after standing for 1 hour and note any separation.

5. If the tank mixture is compatible, any separation observed will be constant in all aliquots taken over the spray cycle.

6. Failure occurs when the sample will not pass through the in-line filter, or spray nozzle. Another indication of failure is the non-uniform separation of the spray aliquots or heavy separation in the spray tank.

The second test used was the shaker test (Illustration 2). The shaker test, an alternate dynamic method, is designed to evaluate a mixed and agitated solution, by passing them through a 100 mesh sieve screen. A system of grading the mixtures was used to reduce the subjectiveness. The only variable is the mesh size of screen employed. The sieve screen's mesh size duplicates the in-line strainer and spray nozzle filters of a tractor's spray rig. Although this method will indicate incompatibility with the clogging of the screen it can not duplicate actual spray conditions.

The test can be graded several ways, but the conclusions are based upon whether the material passed through the screen. The first observation is of the bottle. A 5 point system was used with 1 being the best. The second observation pertains to the screen surface after the sample is poured through. A 3 point system was employed with a 1 being the best. A 3,4 or 5 rating of the bottle or a 3 rating for the screen indicates incompatibility. The rating system is presented in Table 4. The results of the shaker test are listed in Table 6.

Table 4 -- Shaker Test Rating System

Bottle
1 Leaves no film or residue.
2 Light film or residue that rinses out completely.
3 Light film or residue that does not rinse out completely.
4 Heavy film or residue that rinses out completely.
5 Heavy film or residue that does not rinse out completely.

Screen
1. No film or residue.

2. Film or residue that rinses.

3. Film or residue that doesn't rinse out completely.

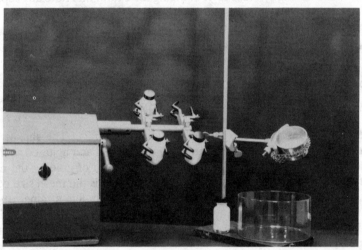

Illustration 2 -- Shaker Compatibility Tester

Compatibility Test #2 -- Shaker Test

Materials

  - Wrist action shaker, Burrell Model or comparable
  - 2 or 4 ounce bottles
  - Carrier of choice
  - 3 inch diameter mesh screens, 50 and 100 mesh size

PROCEDURE

1. Fill appropriate bottles with 50 or 100 ml of carrier in the 2 or 4 ounce bottles respectively.

2. Mark one bottle with no pesticide as the control. In each of the subsequent bottles add the appropriate amount of pesticide. In the final bottle add the total mixture of pesticides.

Note: If more than one pesticide is to be used, add them separately in the following order.

1. Wettable powders
2. Water dispersible granules
3. Flowables
4. Water solubles
5. Emulsifiable concentrates

3. Place the bottles on the wrist action shaker. Set the shaker intensity adjustment to ten.

4. Agitate for 30 minutes, remove the bottles and swirl once or twice to re-suspend any solids.

5. Immediately pour the contents of the bottle through the appropriate mesh screen. If the bottle contains a residue, or the screen has a deposit on it, the bottle and screen may be rinsed with two 50 ml portions of test carrier.

6. Examine the bottle and screen to determine the amount of material that has adhered to the glass or failed to pass through the screen. The presence of material should be verified by using the bottles containing only the components and carrier.

7. If appreciable amounts of material are left in the bottle the sample has failed.

8. Any retention of solid material on the screen is a failure.

The final method used for determining compatibility was the mixing cylinder test (Illustration 3). This static test is a quick method to indicate possible problems. This visual and subjective test introduces questionable reproducibility between analysts. The test requires that all pesticides be screened individually before the combinations are tested. This identifies any pesticide that causes incompatibility.

The test requires a 100 ml mixing cylinders and the carriers of choice. The mixing cylinders are filled to the 80 ml mark with carrier. A 40% slurry of pesticide in 342 ppm water is prepared and 10 ml of this slurry is added to the mixing cylinder. The remaining 10 ml, will be used for the co-herbicide addition. The resulting solution is inverted 10 times and the observations are made at 30 and 60 minute intervals. The degree of separation and redispersion is noted. Incompatibility occurs when the product does not redisperse or will not pass through a 100 mesh screen. The results of the Mixing Cylinder Test are listed in Table 7.

<u>Compatibility Test #3</u> Mixing Cylinder Test

<u>Materials</u>

- 100 ml mixing cylinders
- 50 ml beaker
- Carrier of choice
- 100 mesh sieve screen

<u>Procedure</u>

1. Prepare a 40% slurry of dry flowable to be tested.

2. Fill respective mixing cylinders with 80 ml with the carrier of choice.

3. Add 10 ml of the slurry and fill to 100 ml mark with 342 ppm water.

4. For the combination products add at appropriate levels and fill to the 100 ml mark with 342 ppm water.

Note: If more than one pesticide is to be used, add them separately in the following order.

1. Wettable Powders.
2. Water Dispersible Granules.
3. Flowables.
4. Water Solubles.
5. Emulsifiable Concentrates.

Illustration 3 -- Mixing Cylinder Compatibility Test Equipment

DISCUSSION

LASSO 4E and DUAL 8E were intentionally chosen for this study. Compatibilities in liquid fertilizers and especially with liquid fertilizers and the triazines have been a problem in the field.

Although all three tests methods were used, the best indicator of tank mix compatibility was the Agricultural Laboratory Sprayer (Table 5). The ideal situation was for the filter to remain clear and a uniform pesticide distribution at the spray nozzle. A clogged filter or non-uniform pesticide distribution pattern indicated a compatibility problem.

In all of the Laboratory Sprayer tests one of three conditions occurred. The first condition was for the filter to clog stopping the test. This took place with the commercial dry flowables in Figures 3, 4, 5 and 6. The second condition was the formation of a heavy layer in the spray tank that contained most of the tank mixed pesticides. In Figures 1-6 it was indicated by a continual drop in the pesticide separation percent in the cylinders. This was observed with the commercial dry flowables as well as with the sulfonated naphthalene-formaldehyde condensate atrazine dry flowable. The third condition was a uniform pesticide distribution pattern. All of the dry flowables based on the new dispersants maintained a uniform pesticide distribution pattern.

Table 5 -- Agricultural Laboratory Sprayer Compatibility Test Results

PESTICIDE	FERTERLIZER		
	32-0-0	10-34-0	30-0-0-2
LASSO 4E + ATRAZINE 90DF			
STEPAN	C	C	C
GENERIC	X	X	X
COMMERCIAL	X	X	X
LASSO 4E + CYANAZINE 90DF			
STEPAN	C	C	C
COMMERCIAL	X	X	X
LASSO 4E + LINURON 50DF			
STEPAN	C	C	C
COMMERCIAL	X	X	X
DUAL 8E + ATRAZINE 90DF			
STEPAN	C	C	C
GENERIC	X	X	X
COMMERCIAL	X	X	X
DUAL 8E + CYANAZINE 90DF			
STEPAN	C	C	C
COMMERCIAL	X	X	X
DUAL 8E + LINURON 50DF			
STEPAN	C	C	C
COMMERCIAL	X	X	X

NOTE: C= Compatible, X= Incompatible

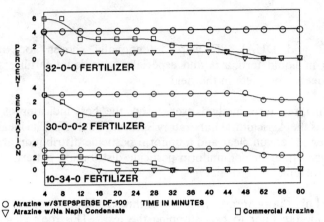

FIG. 1 --- LASSO 4E / ATRAZINE 90DF Spray Compatibility Test

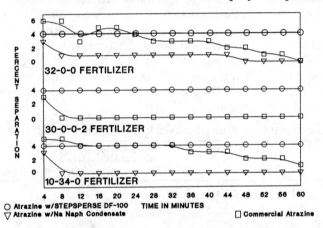

FIG. 2 --- DUAL 8E / ATRAZINE 90DF Spray Compatibility Test

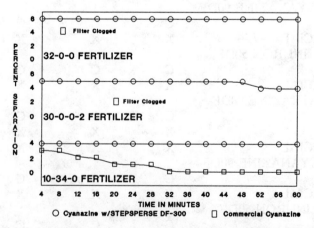

FIG. 3 --- LASSO 4E / CYANAZINE 90DF Spray Compatibility Test

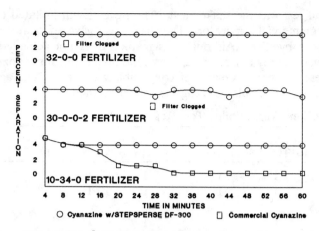

FIG. 4 --- DUAL 8E / CYANAZINE 90DF Spray Compatibility Test

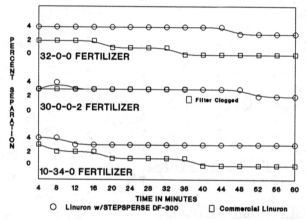

FIG. 5 --- LASSO 4E / LINURON 50DF Spray Compatibility Test

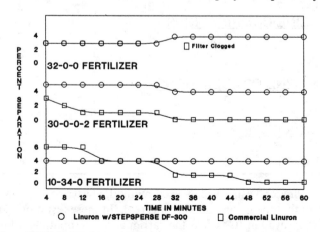

FIG. 6 --- DUAL 8E / LINURON 50DF Spray Compatibility Test

The second test was the shaker test. The sieve screen imitated the in-line filter of the commercial applicator. The shaker test gave adequate results but it was still too subjective. Although this method does not actually spray the solution through a spray nozzle, it duplicates the in-line and spray nozzle filters. This was where the first problems of compatibility occured. The actual field conditions can not be simulated using the shaker method.

Table 6 -- Shaker Compatibility Test Results

Products	32 N		10-34-0		30-0-0-2	
	BOTTLE	SCREEN	BOTTLE	SCREEN	BOTTLE	SCREEN
LASSO EC	1	1	1	1	1	1
DUAL 8E	1	1	1	1	1	1
ATRAZINES						
STEPAN	1	1	1	1	1	1
GENERIC	1	1	2	1	1	1
COMMERCIAL	2	3	2	3	3	2
CYANAZINES						
STEPAN	1	1	1	1	1	1
COMMERCIAL	1	1	1	1	1	1
LINURONS						
STEPAN	1	1	1	1	1	1
COMMERCIAL	1	2	1	2	3	2
LASSO ATRAZINE COMBINATIONS						
STEPAN	3	2	3	3	3	3
GENERIC	4	3	5	3	4	3
COMMERCIAL	4	3	5	3	4	3
LASSO CYANAZINE COMBINATIONS						
STEPAN	2	1	4	1	3	1
COMMERCIAL	2	2	2	2	2	1
LASSO LINURON COMBINATIONS						
STEPAN	4	3	2	3	4	3
COMMERCIAL	2	1	2	1	2	1
DUAL ATRAZINE COMBINATIONS						
STEPAN	2	1	4	3	2	1
GENERIC	2	2	4	3	2	2
COMMERCIAL	4	3	5	3	4	3
DUAL CYANAZINE COMBINATIONS						
STEPAN	2	3	5	3	4	3
COMMERCIAL	2	2	5	3	1	1
DUAL LINURON COMBINATIONS						
STEPAN	2	1	2	1	2	1
COMMERCIAL	4	3	3	2	4	3

The Mixing Cylinder Test, a widely used method, always required additional testing to confirm that the incompatibility was real. In these test it predicted that any combination sprayed in 10-34-0 as the carrier was incompatible, while in fact all the formulations with new dispersant blend were sprayable. This test can only be used as preliminary screen to rule out major incompatibility when the spray mixture becomes a non-flowing suspension.

Table 7 -- Mixing Cylinder Test Results

PESTICIDE	FERTERLIZER		
	32-0-0	10-34-0	30-0-0-2
LASSO 4E + ATRAZINE 90DF			
STEPAN	C	X	C
GENERIC	X	X	X
COMMERCIAL	X	X	X
LASSO 4E + CYANAZINE 90DF			
STEPAN	C	X	C
COMMERCIAL	X	X	X
LASSO 4E + LINURON 50DF			
STEPAN	C	X	X
COMMERCIAL	X	X	X
DUAL 8E + ATRAZINE 90DF			
STEPAN	C	X	C
GENERIC	X	X	X
COMMERCIAL	X	X	X
DUAL 8E + CYANAZINE 90DF			
STEPAN	C	X	C
COMMERCIAL	X	X	X
DUAL 8E + LINURON 50DF			
STEPAN	C	X	C
COMMERCIAL	X	X	X

NOTE: C= Compatible, X= Incompatible

CONCLUSIONS

Dry flowable tank mix compatibility is a problem. The formulator has not up to this time had the tools necessary to correct the problem. These new dispersants will give the formulator the necessary tools and enable him to develop dry flowables with a minimum of tank mix compatibility problems. Results from the laboratory sprayer, shaker and the mixing cylinder indicated that the new dispersants reduced or eliminated incompatibly in all three formulations. These blends, which are new to the industry, allow for the complete manipulation of a formulation as is done with if it was an emulsifiable concentrate. The new dispersant blends offers the formulator many choices. These choices begin an era of new creativity not before available.

ACKNOWLEDGMENTS

The author wishes to acknowledge with appreciation the help of Meredith Potochnic in the preparation of this paper.

REFERENCES

[1] Smith, N.R., "Evaluation of Tank Mix Compatibility Using the Laboratory Sprayer," Pesticide Formulations and Application Systems: 3rd Symposium, ASTM STP 828, American Society for Testing and Materials, Philadelphia, 1983

[2] Ashworth, R. DEB., Henriet, J. and Lovett, J.F., "Analysis of Technical and Formulated Pesticides", CIPAC Handbook, 1970, Volume 1, pp 861-868.

Janet S. Catanach and Sherman W. Hampton

SOLVENT AND SURFACTANT INFLUENCE ON FLASH POINTS OF PESTICIDE FORMULATIONS

REFERENCE: Catanach, J. S. and Hampton, S. W., "Solvent and Surfactant Influence on Flash Points of Pesticide Formulations," Pesticide Formulations and Application Systems: 11th Volume, ASTM STP 1112, Loren E. Bode and David G. Chasin, Eds., American Society for Testing and Materials, Philadelphia, 1992.

ABSTRACT: Intensified regulatory activity from various sources is guiding the consideration of higher flash (lower volatility) materials as preferred components in pesticide formulations. An update on the current issues surrounding storage and packaging of flammable and combustible liquids is presented.

For this paper, the term "components" refers to nonactive materials and includes solvent or solvent blends and surfactants used in the formulations. The purpose of this paper is to study the effect of multiple component solvent systems on flash point. Also, the effect of surfactants in the formulation will be reviewed.

Nonlinearity of flash points when different components are blended for the purpose of improving active solubility, formulation efficacy, evaporation rate, or other parameter is demonstrated.

KEYWORDS: flash point, solvent, surfactant, co-solvent, vapor pressure, regulation

INTRODUCTION

The U. S. government, as well as several states, are calling for reductions in volatile organic emissions and regulating the use of solvents and formulations containing solvents by imposing ceilings on product vapor pressure and reactivity (with nitrous oxide). High flash products and improved packaging materials can improve product safety by reducing fire risk and damage potential.

A generalized Emulsifiable Concentrate (EC) incorporates solvents, solvent-blends, and emulsifiers as components in the

Ms. Catanach is Environmental Affairs Coordinator for hydrocarbon solvents at Exxon Chemical Company, 13501 Katy Freeway, Houston, Texas 77005. Mr. Hampton is a research technician for Exxon Chemical Company, 5200 Bayway Drive, Baytown, Texas 77520.

formulation. The primary solvent is seen as the major contributor to the formulation's flash point, but other components also help determine the EC's flash point.

Reformulation efforts are underway to produce pesticide products without fire points. However, some active ingredients are chemically sensitive, and may decompose in storage if not formulated with organics. In addition, formulations containing water allow less efficient coverage of the active for some applications since droplet size is not properly controlled [1].

Solvents, co-solvents, and surfactants are therefore necessary components in agricultural formulations. Until formulation technology finds alternatives to these materials, high flash, low volatility products are recommended to decrease fire risk and atmospheric reaction.

REGULATIONS, SAFETY, AND CONTROL/PREVENTION

Specific regulatory activity aimed at pesticide formulation components include the Clean Air Act of 1990 [2], U.S. Department of Transportation codes [3], state rules such as those enacted or proposed by New Jersey [1], California [4], and Minnesota (tax on federally regulated pesticides and flammables) [5], cities such as New York City's Fire Department [3], and business and interest groups such as the American Iron and Steel Institute [6] [7], and Factory Mutual [8] [9].

Most of the legislation is aimed at reducing volatile organic content (VOC) emissions prioritizing emittors based on reactivity (since not all VOC react the same in sunlight with nitrous oxide to form ozone). As the issue of VOC reactivity is new, most regulation of solvents is achieved by imposing restrictions on the product's vapor pressure. Vapor pressure limitations vary-- 1330 Pa (1 mmHg) at 20 degree C for New Jersey; 133 Pa (0.1 mmHg) at 20 degree C for New York; and in California, the proposed non-regulated limit is 13.3 Pa (0.01 mmHg) at 20 degree C, with regulated vapor pressures ranging from 13.3 Pa (0.01 mmHg) to 2660 Pa (2 mmHg).

These vapor pressure limitations directly relate to any component's flash point since flash point is a vapor/temperature relationship, and the fraction volatilized is dependent on a product's vapor pressure. For example, lowering the vapor pressure proportionately raises the flash point since it takes a higher temperature to effect volatilization of enough of a low vapor pressure material to form a mixture which will support combustion. Therefore, as regulation supports low vapor pressure formulations and components, it necessitates formulations be higher flash and also contain higher flash components.

Other agencies and groups directly address the issue of flash point to enhance product safety. The U. S. Department of Transportation has signaled a move to International Shipping Standards and Instructions effective January 1, 1991. These standards will change labeling of products having flash points of

37.8-60 degree C (100-140 degree F) from combustible to flammable and flash points of products greater than 60.5 degree C from combustible to non-regulated. Not associated with transportation regulation is the additional storage volume limitation for products with flash points under 60 degree C (140 degree F) in unprotected indoor storage areas. Also, mixing of materials of different flash points implements an additive volume restriction [10].

The National Fire Protection Agency (NFPA), Factory Mutual, New York City and others have proposed restrictions or bans on plastic containers such as polyethylene but currently applying to all thermoplastic substances. The limitation cites flammable liquids due to test information from the American Iron and Steel Institute and other sources indicating higher risk potential associated with storage of flammables in plastic containers. In this context, flammable products are those currently classified as Class I and II liquids having flash points below 60 degree C (140 degree F). The amendment to NFPA Code 30 restricting Class I and II liquids in plastic containers for storage in general purpose warehouses has been moved from September 1, 1990 to September 1, 1991. When final, this amendment will require plastic container storage in inside rooms or liquid warehouses.

Plastic container storage requirements (and associated cost) will impact pesticide suppliers since plastic containers are preferred for pesticides due to added safety (lower breakage relative to glass) and improved economics (plastics are less expensive to make and are lighter than metal and glass therefore reducing shipping costs). Another issue supporting plastic containment over metal has been the subject of BLEVE (Boiling Liquid Expanding Vapor Explosion). This involves flammable liquid containment in metal. However, recent data indicates smaller cans are not subject to BLEVE when under common control devices such as sprinklers. Typically smaller cans have vents, lids that pop off or gaskets that burn out preventing explosion. Therefore, the plastic vs. metal storage issue relates to container size of less than five gallons. Additional data on one gallon and lesser volumes shows plastic containers may melt releasing all liquid, giving off smoke, thereby creating a situation which cannot be controlled by sprinklers of the type and configuration in a general purpose warehouse.

Therefore, if plastic containers are to remain viable packaging materials for pesticides, the formulations and also formulation components must support a flash in excess of 60 degree C (140 degree F).

The plastic container issue has exemptions for containers less than 16 ounces if holding water-miscible liquid present in greater than 50 volume percent or for products having a composition of not more than 50 volume percent water-miscible liquid with remainder of the material being nonflammable.

FLASH POINT

Regulation and safety requirements spurring utilization of higher flash point products prompts discussion of flash point testing.

For the flash point ranges identified in NFPA Code 30 on Flammable and Combustible Liquids, several methods are cited. Those relative to pesticide formulation and components are ASTM D56, Standard Method of Test for Flash Point by the Tag Closed Tester and ASTM D93, Standard Method of Test for Flash Point by the Pensky-Martens Closed Tester [11, 12]. The former method is applicable to products having a viscosity less than 5.5×10^{-6} m^2/s (5.5 cSt) at 40 degree C or less than 9.5×10^{-6} m^2/s (9.5 cSt) at 25 degree C and a flash point less than 93 degree C unless products in this range contain suspended solids or form a surface film. If the latter is true or if products have a viscosity greater than 5.5×10^{-6} m^2/s (5.5 cSt) at 40 degree C, the recommended test method is Pensky-Martens.

The precision for both methods has been tested and differs for varying flash points. In addition, due to the difference in the two procedures, the same flash point result should not be anticipated on any given product. For example, a product containing alkylbenzenes was tested by both methods receiving a value of 44.0 degree C (111 degree F) using ASTM D56 and a value of 40.5 degree C (105 degree F) using ASTM D93. The primary reason for the lower flash point obtained using ASTM D93 is the application of stirring in this method. Therefore, a comparison of flash points is only valid when the same method is used throughout the comparison.

FLASH POINTS OF PESTICIDE FORMULATION COMPONENTS

The flash point of an emulsifiable concentrate, in general, is dominated by the flash point of the solvent contained in the formulation. The term "solvent" however, may have many sources as is about to be explained. For the purpose of discussion on the impact of the EC's flash point from components, a generalized formulation is given as 45% (range 40-60%) solvent, 5-6% (range 2-12%) emulsifier, and 50% (range 40-60%) active ingredient [13]. Another assumption to be stated for the purpose of further examination of flash point is the composition of the surfactant portion of the EC.

Generic emulsifier formulations may be comprised either of equal contributions of a sulfonate, an ethoxylated castor oil and a nonyl phenol block co-polymer or a 70/30 blend of a single nonionic such as a nonyl phenol and an anionic such as a sulfonate. As indicated, these packages are present at levels of 2-12% or at an average of 5-6%. Also, these emulsifiers are generally 60-80% active meaning 40% of the 5-6% is solvent. Therefore, the contributing portion of the emulsifier to flash point may be as much as 2% or more of the total emulsifiable concentrate formulation.

The effect on flash point from the components is tested by excluding the concentration of the EC attributed to the active ingredient and normalizing the other components to 100%. The effect of solvent is analyzed first followed by the effect on flash point with the addition of the emulsifier. Also, since single solvent systems are not always used, typical binary solvent systems are discussed.

Table 1 shows flash points of some commonly used solvents and demonstrates the variety (e.g. compositional differences) of products that may be used as well as the wide range in flash points of these products. The table also shows the effect on flash point when a co-solvent is used. This effect on flash point is similar to what may occur for an EC when multiple solvents are used or when emulsifiers containing solvent are used. The comparison showing high concentration (8-10%) of co-solvent is consistent with the effect expected for binary solvent systems. The lesser concentration (4-5%) is the expected effect from solvent-containing-emulsifiers (if the impact is normalized).

The effect of the co-solvent may be minimal to significant depending on the flash point of the primary solvent. That is, all hydrocarbon solvents' flash points are affected by the addition of small concentrations of a low flash material such as methanol. The impact from higher flash products such as hexanol, methylpyrrolidone-2(N) (NMP), or γ-butyrolactone, have less effect when added as a co-solvent. For example, when a xylene range aromatic solvent (flash point approximately 44.0 degree C) is mixed with a small amount of a high flash product, the flash point may be reduced a few degrees or not at all. The impact (i.e., flash point depression) is greater if this same product is added to a C10 alkylbenzene or a C11 alkylnaphthalene. Test data indicate a flash point may be reduced as much as 22.5 degree C (34 degree F).

The theory behind this effect may be in line with Raoult's Law which discusses changes in a material's vapor pressure (by adding a high vapor pressure product to a relatively low vapor pressure product) thereby increasing the frequency that molecules escape from the surface of a liquid (accelerating the rate of product vaporization to form, at a faster relative rate, a flammable air/vapor mixture which will flash if ignited) [14]. Therefore, if regulation or safety requirements dictate 60 degree C or 93 degree C minimums, flash points of co-solvents used in formulations will have to be scrutinized against the flash point of the primary solvent to insure the flash point of the total formulation meets the requirement.

Additional to actual testing for flash points of such mixtures, calculations and computer models are available for predicting flash points of solvent mixtures [15].

SURFACTANT EFFECT

The emulsifier packages used for this demonstration contain some level of solvent. From the flash point of the emulsifier package, the surfactant appears to contain a solvent having a flash point similar to that of a C10 alkylbenzene (see Table 2). Therefore, a minimal flash point reduction is witnessed when the surfactant is added to the C9 aromatic or C10 alkylbenzene. As with the previous demonstration on co-solvent effect, however, the flash point of the mixture of surfactants and C11 alkylnaphthalenes is significantly and statistically different from the flash point of the C11 alkylnaphthalene alone. Therefore, if the goal is to achieve a minimum flash point for the EC of 93 degree C (200 degree F), both

TABLE 1 -- Flash Points: Solvents and Solvent Blends

	Dominant Solvent		
	C9 Xylene Range Aromatic	C10 Alkyl-benzenes	C11 Alkyl-naphthalenes
Flash Point, degree C (F)*	44.0 (111)	65.5 (150)	99.0 (210)
Flash Point After Addition Of Following Percent Co-solvent, degree C (F)			
4% Methanol	13.0 (56)	16.0 (61)	16.0 (61)
8% Methanol	12.0 (54)	14.0 (57)	14.5 (58)
[Methanol Flash Point = 10.5 (51)]			
4% Hexanol	42.0 (108)	61.5 (143)	80.0 (176)
8% Hexanol	41.0 (106)	59.0 (138)	76.5 (170)
[Hexanol Flash Point = 58.0 (136)]			
5% NMP	42.0 (108)	65.5 (150)	101.0 (214)
10% NMP	42.0 (108)	66.0 (151)	100.0 (212)
[NMP Flash Point = 83.0 (181)]			
5% ɣ-Butyrolactone	45.5 (114)	69.0 (156)	96.5 (206)
10% ɣ-Butyrolactone	42.0 (108)	65.0 (149)	93.5 (200)
[ɣ-Butyrolactone Flash Point = 99.5 (211)]			

*   Flash Point Test Methods used when following solvent
     alone or dominant in blend:
     C9 Xylene Range Aromatic - ASTM D56
     C10 Alkylbenzenes - ASTM D56
     C11 Alkylnaphthalenes - ASTM D93

TABLE 2 -- Flash Points: Solvents and Solvent/Surfactant Blends

	Dominant Solvent		
	C9 Xylene Range Aromatic	C10 Alkyl- benzenes	C11 Alkyl- naphthalenes
Flash Point, degree C (F)*	44.0 (111)	65.5 (150)	99.0 (210)
Flash Point After Addition Of 10% Trinary Emulsifier Package, degree C (F)	44.0 (111)	63.5 (146)	91.0 (196)

[Trinary Emulsifier Package Flash Point = 61.5 (143)]

Flash Point After Addition Of 10% 70/30 Emulsifier Blend, degree C (F)	41.0 (106)	63.5 (146)	92.0 (198)

[70/30 Emulsifier Blend Flash Point = 68.0 (154)]

*   Flash Point Test Methods used when following solvent alone or dominant in blend:
     C9 Xylene Range Aromatic - ASTM D56
     C10 Alkylbenzenes - ASTM D56
     C11 Alkylnaphthalenes - ASTM D93

co-solvent and surfactant package flash points must be reviewed prior to formulation.

CONCLUSION

National, state, and urban regulation as well as increased safety awareness and requirements are imposing the need for high flash (low volatility, low vapor pressure) products for use in agricultural formulations.

Products exist which may meet these requirements, but it must be recognized that all components in a formulation contribute to the product's ultimate flash point. The volatility as well as functionality of co-solvents or surfactants must be reviewed prior to formulation to insure compliance with regulation and safety requirements.

REFERENCES

[1] Namnath, J.S., et al., "A Study of Volatile Organic Compounds (VOCs) and Photochemically Reactive Organic Compounds (PROCs) in Relation to Household Pesticide Formulations," Pesticide Formulations and Applications Systems: International Aspects, 9th Volume, ASTM STP 1036, James L. Hazen and David A. Hovde, Eds, American Society for Testing and Materials, Philadelphia, 1989.

[2] Environmental Protection Agency, Federal Register, Part XXII, Semi-annual Regulatory Agenda, April 23, 1990.

[3] Chemical Specialties Manufacturers Association, Executive Newswatch, "New York Fire Department Rule to Regulate Plastic Containers For Flammable Liquids," Volume 38, No. 14, October 1, 1990.

[4] CAPCOA Pesticides Solvents Task Force, Meeting Number 4 Summary, June 22, 1990

[5] Chemical Specialties Manufacturers Association, Legislative Reporter, August, 1990.

[6] Committee of American Iron and Steel Institute, CN-03, Fire Safety: Flammable and Combustible Liquid Container Storage, Tin Mill Products Producers, 1133 15th Street NW, Washington, D.C. 20005-2701.

[7] Hughes Associates, Inc., for American Iron and Steel Institute, Fire Tests of Steel and Plastic Containers of Paint Thinner, CN-06, April 16, 1987.

[8] Factory Mutual Engineering Corp., "Flammable Liquids In Drums and Smaller Containers," Loss Prevention Data 7-29, September, 1989.

[9] Factory Mutual System, Record, "Flammable Liquids in Plastic Containers," Jan./Feb. 1987.

[10] National Fire Protection Agency, NFPA 30 Flammable and Combustible Liquids Code, 1987 Edition.

[11] Annual Book of ASTM Standards, Volume 05.01, ASTM D56 Standard Test Method for Flash Point by Tag Closed Tester.

[12] Annual Book of ASTM Standards, Volume 05.01, ASTM D93 Standard Test method for Flash Point by Pensky-Martens Closed Tester.

[13] Consultation with personnel in Stepan, Witco, Dow-Elanco, and Monsanto.

[14] General Chemistry With Qualitative Analysis, Fourth Edition, D. C. Heath and Company, Chapter 12, pgs. 257-258.

[15] American Chemical Society, I&EC Fundamentals, "Flash Points of Flammable Liquid Mixtures using UNIFAC", 1982, 21, 186.

R. Scott Tann, Paul D. Berger, Christie H. Berger

APPLICATIONS OF DYNAMIC SURFACE TENSION TO ADJUVANTS AND
EMULSION SYSTEMS

REFERENCE: Tann, R. S., Berger, P. D., and Berger,
C. H., "Applications of Dynamic Surface Tension to
Adjuvants and Emulsion Systems," Pesticide Formula-
tions and Application Systems: 11th Volume, ASTM
STP 1112, Loren E. Bode and David G. Chasin, Eds.,
American Society for Testing and Materials,
Philadelphia, 1992.

ABSTRACT: Dynamic surface tension (DST) measurements
were obtained using the Maximum Bubble Pressure
Technique for various adjuvant and emulsion systems.
These results were compared with traditional test
methods for adjuvants and emulsion systems.
    The adjuvants were compared to: 1.) Drave's
wetting times, 2.) Static surface tension 3.) Static
contact angle. The relationships between these
properties and dynamic surface tensions at various
frequencies were studied. The emulsion system
studied contained an anionic surfactant and a
nonionic surfactant in a 4 #/gallon pesticide
formulation. The resulting emulsions were measured
for dynamic surface tension at varying frequencies
and compared to traditional emulsion results.
    This work demonstrates the application of dynamic
surface tension to the study of adjuvants and
emulsion stability.

KEYWORDS: surface tension, dynamic surface tension,
adjuvants, emulsion stability, wetting agents

Mr. Tann is a R&D chemist with the Agricultural group of
Witco Inc. . Ms. Berger is an R&D chemist with the
Oilfield product group for Witco Inc.. Mr. Berger is the
research and development director for Witco Inc.. All
three authors can be found at 3200 Brookfield St.,
Houston, TX 77045.

The role of the adjuvant in the agricultural marketplace has been increasing for many years. With the economic and environmental concerns associated with the use of pesticides, adjuvants have found a place in the agricultural scene. The term adjuvant has become synonymous with the following products: wetting agents, spreaders, stickers, foaming agents, defoamers, crop oil emulsifiers, humectants, compatibility agents, and soil penetrants to name a few. One might admit the term adjuvant applies to every product used in a spray tank which does not contain a toxicant. The use of these adjuvants can effectively reduce the amount of pesticide needed to produce the desired effect thus reducing some economic concerns of the applicator. The use of some of these adjuvants can aid in deposition of the pesticide on the target thus reducing the environmental impact of the pesticide. Many current producers of agrichemicals have incorporated certain adjuvants into their formulations. Still others have modified their product booklets or label instructions to indicate the proper use level of adjuvants to employ with their products. With the proliferation of the use of agricultural adjuvants many different testing and evaluation procedures have become common. Formulators are presently evaluating adjuvants based on their end use with differing methodology. In this paper we present a technique for evaluating many of these adjuvants currently in use. This technique is the Maximum Bubble Pressure (MBP) technique for obtaining Dynamic Surface Tension (DST). Application of this technique to adjuvant products will aid in our knowledge of these products and their performance. It will be shown that DST directly correlates with two present test methods in use to evaluate adjuvants. We will also show one of the present measurement techniques fails to accurately reflect the performance of the adjuvant. It will be proposed that the DST method be used as a tool in the evaluation and quality control of adjuvants.

Another application of this technique was found in the area of stability of traditional oil in water macroemulsions. Comparison testing of certain herbicide and insecticide formulations using traditional emulsion separation scores and DST measurements were obtained to investigate the relationship of the two test methods. The comparison testing was performed in a range of water hardnesses as well as a range of temperatures. By evaluating the pesticide emulsion in both testing protocol certain correlations and applications were determined.

This paper describes the findings of a study to investigate the application of the MBP technique for DST to adjuvants and emulsion systems. The relationship of the DST to present and proposed methodology for evaluating adjuvants will be discussed. The use of the technique to evaluate emulsions also will be discussed.

APPARATUS AND PROCEDURE

Contact angles for the adjuvants were obtained by using a Rame-Hart Inc. Goniometer model # 100-00115. The contact angle of a sessile drop was obtained on a microscope slide covered with parafilm. The parafilm was used to simulate a leaf surface and correlated directly to a leaf sample when tested. The drop was placed onto the surface using a calibrated syringe with a 22 gauge needle. Distilled water and benzene on paraffin wax were used as standards to insure the accuracy of the findings. Duplicate samples of each standard were evaluated and the resulting precision determined to be ± 1 degree. The values were obtained at 25 degrees Celsius as a 0.1 % wt/wt solution of adjuvant to deionized water respectively. The angle of measurement was the acute angle. Contact angle measurements currently are used in the evaluation of spreaders and stickers. The lower the contact angle the better chance the pesticide has at sticking to the plant. As the contact angle of the droplet increases so too does the chance of runoff of the pesticide from the target surface [1].

Wetting Times of the adjuvants were obtained by the ASTM standard method: D2281-68 [2]. Duplicate trials were performed and a resulting error of ± 2 seconds was obtained. Solutions were prepared at 0.1 % wt/wt in deionized water and allowed to equilibrate for 12 hours prior to testing. Wetting times are the currently accepted method of evaluating wetting agents and activators. The smaller wetting times indicate better field performance for the adjuvant. Wetting times for adjuvants have generally been considered acceptable under 100 seconds.

Static Surface Tension measurement of the adjuvants were determined by the Du Nouy ring method [3]. Solutions were prepared at 0.1% wt/wt in deionized water and allowed to equilibrate for 12 hours prior to measurement. By determining the surface tension between deionized water and air and comparing the result to the literature value of 72.3 dyne/cm at 25 degrees Celsius the accuracy of the method was established. This resulted in a sensitivity of the method of ± 0.02 dyne /cm. Static surface tension measurements have long been considered the method of choice for evaluating surface active agents. The lower the surface tension the better in most applications of adjuvants. DST measurements for the adjuvant products was obtained using the Maximum Bubble Pressure Technique (MBP) [3,4]. Solutions from the static surface tension measurements were used in this study.

DST measurements for the emulsion preparations were obtained using the (MBP) technique with some deviation[4]. The opacity of the resulting emulsion forced the use of a pressure transducer for the detection of the bubble.

Prior references use a photodiode to detect the presence of the bubble. Measurements were made in different water hardnesses (34 ppm, 342 ppm and 1000 ppm) as was necessary to notice the effect of hardness. By comparison of obtained values between air and deionized water the sensitivity of the method was determined to be $\pm$ 0.02 dyne/cm error. The emulsions were prepared at 5% in the different hardness waters (34 ppm, 342 ppm and 1000 ppm) for the respective tests. The emulsion was prepared and poured into the tensiometer chamber and necessary air flow was allowed to bubble through the emulsion.

Separation was ranked by an arbitrary emulsion score. Emulsion separation was assigned a value by the following scale: <u>Separation Rating Scale</u>

    1 = No visible separation
    2 = Trace cream
    3 = Light cream
    4 = Moderate cream
    5 = Heavy cream
    6 = Dispersed oil
    7 = Free oil

Emulsion scores were prepared by multiplying the separation value by the milliliters of separation noted. Five milliliters of the emulsifiable concentrates were introduced into 100 ml graduated cylinders containing ninety-five milliliters of the test water. The 100 ml cylinders were graduated in 1 ml divisions to facilitate low cream readings. Spontaneity was noted as the emulsifiable concentrate was placed in the water using a repeatable pipet. Cream readings and separation ratings were obtained at 1/2 hour, 1 hour, 2 hour, and 24 hour intervals. Emulsion scores were then plotted against the anionic/nonionic ratio. The experimental error for this method was determined to be $\pm$ 0.5 ml by duplicate trials of the emulsifiable concentrates in the test medium.

The emulsifiable concentrates used in this study consisted of a herbicide and an insecticide. The herbicide chosen was a 4 #/gallon formulation of Isooctyl Ester of 2,4-Dichlorophenoxyacetic Acid (IOE/2,4-D) in Exxon Aromatic 150. The insecticide chosen was a 4 #/gallon formulation of Chlorpyrifos in Exxon Aromatic 150. Emulsifier levels were established at 6% for the insecticide formulation and 5% for the herbicide formulation. The anionic emulsifier differed between the herbicide and the insecticide formulation. The IOE/2,4-D formulation utilized an amine C-12 alkylarylsulfonate whereas the Chlorpyrifos utilized a calcium C-12 alkylarylsulfonate. Both formulations made use of the same nonionic namely a castor oil ethoxylate. Separate solutions of the pesticides containing the nonionic and anionic emulsifiers were prepared. These solutions were blended at the test ratio prior to addition into the test water.

RESULTS AND DISCUSSION

Adjuvants

DST measurements were obtained for a representative wetting agent. The wetting agent used was a blended formulation known as Adsee 100. Contact angle, Drave's wetting time and static surface tension were directly compared to the DST. Percentage by weight of the wetting agent in deionized water was varied to produce Table 1. The results in Table 1 indicate the direct correlation between Contact Angle, Drave's wetting time and DST values at 1 sec. The static surface tension measurements however do not correlate with the other test results.

TABLE 1 -- Adsee 100 Surface Tension and Wetting

Wt. %	Contact Angle degrees	Draves Wetting sec.	Static S.T. dyne/cm	D.S.T. @ 1 sec dyne/cm
1.00	36.0	9	31.30	37.0
0.10	47.0	107	31.68	62.0
0.01	62.0	>300	31.68	71.0

It was hypothesized that as the wetting agent concentration is decreased the surface properties should decrease as well. All test results support this hypothesis with the exception of the static surface tension. Little or no change in static surface tension is observed at the three concentrations listed. The data also shows all the concentrations of Adsee 100 are at or above the CMC. CMC refers to the critical micelle concentration at which the static surface tension becomes constant due to the saturation of the surface with surfactant. All other properties continue to change except the static surface tension. This product is currently used at the 0.5% to 1.0% rate in the field. The data from Table 1 suggests the product should be used at the present rates to obtain the optimum results.

Table 2 shows a series of sulfosuccinates which have been found to be good wetting agents or dispersing agents.

TABLE 2 -- Emcol Sulfosuccinates

Product	Sulfosuccinate Description	Applications Base
4500	Di 2-ethylhexyl	wetting,grinding
4910	Polypropoxy,C12-15 alkyl	pigment dispersant
4300	Monolaureth	oil wetting, grinding
4161	Monooleamido	pigment dispersant
K-8300	Monoalkanolamide	pigment suspension

The sulfosuccinates were compared in their surface
activity by four methods. Contact angle, static surface

tension, Drave's wetting time and DST were determined. The
results are listed in Table 3. The data has been arranged
in order of increasing contact angles.

TABLE 3 -- Emcol Sulfosuccinates - Wetting Properties

Product 0.10 %	Contact Angle degrees	Drave's Wetting sec.	Static S.T. dyne/cm	D.S.T. @ 1 sec. dyne/cm
4500	25.0	< 1	34.5	49.5
4910	27.0	10	26.8	63.6
4300	34.0	18	28.1	69.8
4161L	35.0	90	34.0	70.2
K-8300	62.0	90	42.1	70.3

As the data suggests all surface properties correlate to
each other with the exception of the static surface
tension values. This data would suggest the method of
evaluating surfactants by simply checking static surface
tension may not accurately reflect the surface properties
of the product.

The data presented in Table 4 represents various
adjuvants and their surface properties. Different types of
adjuvants were obtained to compare DST to the contact
angle, static surface tension, and wetting time.

TABLE 4 -- Common Adjuvants Surface Properties @ 0.10 %

Function	Contact Angle degrees	Drave's Wetting sec.	Static S.T. dyne/cm	D.S.T. @ 1 sec dyne/cm
Comp. Agent	39.0	70	32.0	51.0
Activator	30.0	50	28.0	66.0
Spreader/Sticker	35.0	50	31.0	67.0
Crop Oil Emuls.	42.0	300	31.0	64.0
Foaming Agent	40.0	90	28.0	63.0
Soil Penetrant	21.0	73	31.0	66.0
Wetting Agent	21.5	8	29.0	44.0

A close look at the data suggests the use of DST measurements accurately reflects the wetting properties of the group of adjuvants. The data also suggests the need for other measurement techniques to evaluate the products not considered as wetting agents.

Emulsion Stability

    Many attempts have been undertaken to determine the stability of emulsions with methodology other than separation time. These methods have included measurement of zeta potential, particle sizing techniques and centrifugal techniques to name a few. Traditional emulsion testing involves the use of graduated cylinders and measuring the cream separation as related to time. In this paper we have attempted to use DST as a tool to predict emulsion stability. Figure 1 shows the results of determining the DST of an aqueous pesticide emulsion. The pesticide in this particular example was a 4 #/gal Chlorpyrifos formulation. This figure shows the effect of bubble frequency vs. the measured DST. These results were obtained in soft (34 ppm) water.

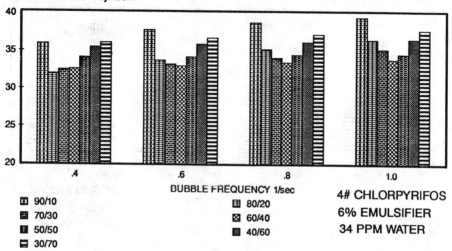

FIG. 1 -- Surface Tensions vs. Frequency

As the figure indicates, at all frequencies listed the
minimum surface tension is obtained at a blend between
70/30 and 60/40 anionic/nonionic respectively. The anionic
emulsifier in this example was a calcium C-12
alkylarylsulfonate. The nonionic emulsifier was a castor
oil ethoxylate. Missing values indicate the lack of
stability in the emulsion during the measurement period.
As the bubble frequency is increased the detection becomes
more difficult. The point of note in this figure is the
trend within each of the frequency groups. Although the
surface tension values change the minimum point in the
curve within the frequency group remained consistent.

    Figure 2 represents the same emulsifiable concentrate
as in Figure 1. The difference between the two figures is
the water hardness. Figure 2 reflects values in hard (1000
ppm) water. It is apparent from Figure 2 that the optimum
ratio of anionic to nonionic is found at 70/30. The
anionic emulsifier and nonionic emulsifier is the same as
in figure 1. As in the previous figure the minimum of the
frequency group occurs in the same ratio.

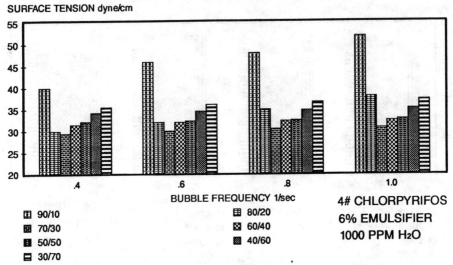

FIG. 2 -- Surface Tensions vs. Frequency

Figure 3 reflects the effect of hardness on the Chlorpyrifos emulsifiable concentrate. A bubble frequency of 1 bubble/second was chosen for the comparison.

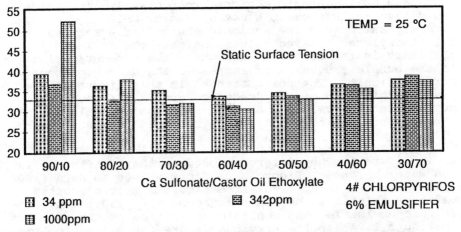

FIG 3 -- D.S.T. @ 1 sec. - Effect of Water Hardness

The data indicates the best possible emulsion stability in all water hardnesses occurs at an anionic to nonionic ratio between 70/30 and 60/40. The data reflects the obvious insensitive nature of the use of static surface tension measurement techniques. The static surface tension fails to differentiate between blends or water hardnesses Traditional emulsions are designed to peak in the 342 ppm water. In Figure 3 the low point occurs at the 342 ppm water as the blend of anionic to nonionic ratio approaches optimum.

Figure 4 reflects the data obtained from traditional emulsion separation methodology. This data represents emulsion scores obtained after 24 hours of separation time.

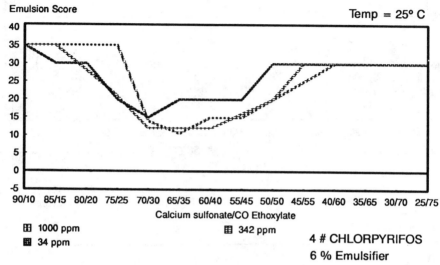

FIG. 4 -- Emulsion Stability - Effect of Hardness

The data in Figure 4 indicates the best emulsion stability occurred at the anionic to nonionic ratio between 70/30 and 60/40. As in Figure 3 the low point reflects the most stable formulation. These results correlate directly with the results obtained in the DST technique. The optimum blend of anionic to nonionic occurred in the same range for both methods. As in the DST measurements the emulsion peak in the water hardnesses occurs at 342 ppm when the optimum ratio of anionic to nonionic ratio is achieved.

The emulsions were evaluated at two different temperatures to determine the effect of temperature on the stability of the systems. The two temperatures chosen were room temperature and 50 degrees Celsius. In order to observe the effect more efficiently the use of 342 ppm water was chosen. Results of the DST are shown in Figure 5. The use of the frequency of 1 bubble/second was again chosen to illustrate the effect.

Figure 6 represents the emulsion scores found at the same temperature used in the DST measurements. The data failed to show any correlation between the two measurement methods. Original theory had predicted that as the emulsion temperature was raised the system would require higher concentrations of nonionic. The emulsion score data found in Figure 6 agreed with the original theory however the findings of the DST measurements failed to agree with the theory.

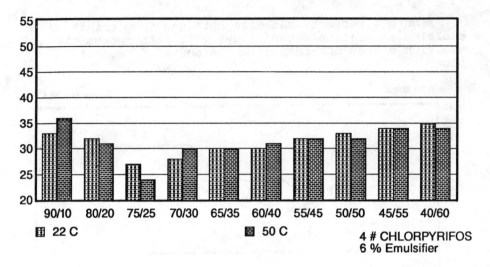

FIG 5 -- D.S.T. @ 1 sec.- Effect of Temperature

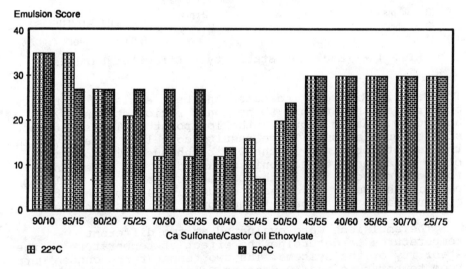

FIG. 6 -- Emulsion Stability - Effect of Temperature

The IOE/2,4-D emulsifiable concentrate utilized a different sulfonate than the Chlorpyrifos emulsifiable concentrate. An amine C-12 alkylarylsulfonate was chosen for the IOE/2,4-D system due to the incompatibility of IOE/2,4-D with calcium salts of sulfonates. This selection also allowed use of a anionic which will more readily dissociate in an emulsion. By utilizing this anionic it was postulated the system would be more selective to water hardness than the calcium sulfonate.

The IOE/2,4-D system was emulsified and the DST measurements were obtained. Figure 7 shows the results of these measurements in soft (34 ppm) water.

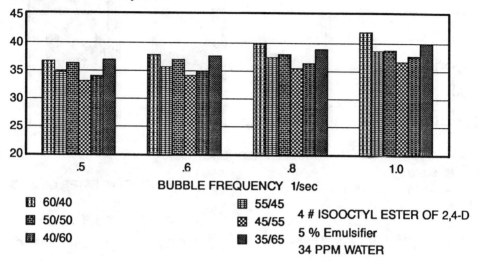

FIG. 7 -- Surface Tensions vs. Frequency in 34 ppm water

From the graph the optimum blend for 34 ppm water was
determined to be 45/55 anionic to nonionic respectively.
As with the previous emulsifiable concentrate
(Chlorpyrifos) the individual groups contain a minimum in
the curves. This minimum occurs at the same ratio of
anionic to nonionic at all frequencies. Figure 8 shows
the results of the DST measurements in hard (1000 ppm)
water. From Figure 8 the optimum ratio of anionic to
nonionic occurs at the 45/55 ratio of anionic to nonionic
respectively. Once again the minimum in all the groups of
frequency data occurs at the same ratio. This data
indicates the IOE/2,4-D emulsion performance is not
dependent on water hardness. Figure 9 reflects the
similar trend as the previous two graphs. The minimum
surface tension occurs at 45/55 ratio of anionic to
nonionic. The bubble frequency was chosen at 1 bubble/sec
to illustrate the effect due to hardness. The IOE/2,4-D
system also was evaluated using separation methodology.
The findings of this test are shown in Figure 10. The
effect of the amine sulfonate becomes obvious in Figure
10. As the emulsion is shifted to harder water the peak
ratio also shifts. The peak ratio was determined to be:
55/45 in 34 ppm water, 50/50 in 342 water and 45/55 in
1000 ppm anionic/nonionic respectively. This would
indicate the need for another surfactant in this system to
increase the span of the emulsifier into all three waters.
This method showed a distinct effect due to hardness with
the amine sulfonate as the anionic.

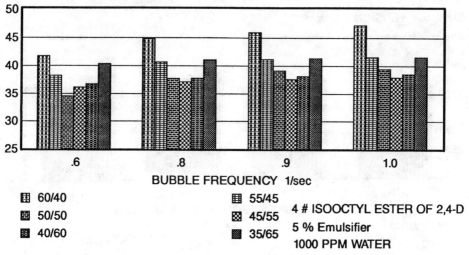

FIG. 8 -- Surface Tensions vs. Frequency in 1000 ppm water

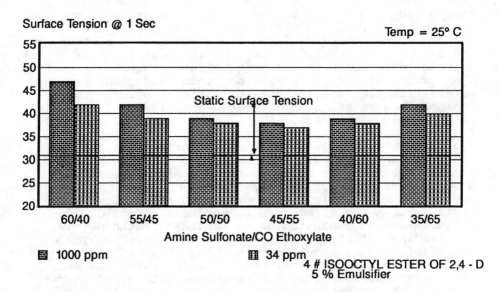

FIG. 9 -- D.S.T. @ 1sec. - Effect of Hardness

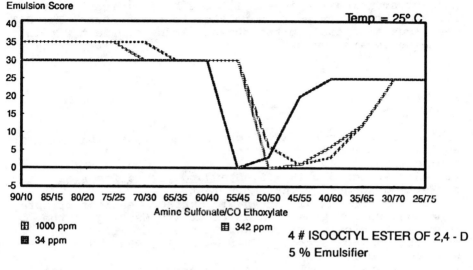

FIG. 10 -- Emulsion Stability - Effect of Hardness

CONCLUSIONS

Dynamic surface tension (DST) is a reliable tool in the evaluation of adjuvants. DST accurately reflects the surface properties of all the adjuvants studied. The use of the DST technique in the evaluation of wetting agents is superior to the present use of static surface tension measurement techniques. A direct correlation was found in the relationship between contact angle, wetting time and DST. DST measurements do not reflect the surface properties of all adjuvants. This study would indicate this technique is a reliable tool for time dependent adjuvant evaluation. Time dependent adjuvant evaluation refers to the dynamic surface properties of the adjuvant. In this study the DST values did not reflect the surface properties of the compatibility agents, soil penetrants, and crop oil emulsifiers. DST technique should be considered as a new tool in the understanding of surface chemistry of adjuvants.

The use of DST measurements in emulsion stability evaluation gives a better understanding of the surface properties at work in the emulsion system. Although DST values change at different frequencies the relationship of anionic to nonionic blend ratios stay consistent throughout each individual frequency. The optimum blend is definable within each frequency group. Static surface tension measurements do not accurately reflect the surface properties of emulsions. No noticeable change was noted in the static surface tension at different hardnesses or temperatures. The DST technique can be used as a tool to predict emulsion stability. It does have limitations for emulsion stability as this relates to temperature. The DST technique also has limitations on hardness effect within a amine sulfonate system.

REFERENCES

[1] Gould, Robert F., Ed., Contact Angle, wettability and adhesion, ACS Advances in Chemistry Series # 43 publication, 1964.
[2] ASTM Standard D2281-68, Standard Method for Evaluation of Wetting Agents by the Skein Test
[3] Berger, P., et al., "Dynamic Surface Tensions of Spray Tank Adjuvants," presented at the 194th Meeting of the American Chemical Society, New Orleans, LA, 1987; available as ACS Symposium Series #137, 1988 ACS Washington D.C.   Chapter 13.
[4] Berger, P., et al., "Dynamic Surface and Interfacial Properties," presented at the 11th ASTM Symposium on Pesticide Formulations and Application Systems, San Antonio, TX.

Deborah R. Fravel and Jack A. Lewis

PRODUCTION, FORMULATION AND DELIVERY OF BENEFICIAL MICROBES FOR
BIOCONTROL OF PLANT PATHOGENS

REFERENCE: Fravel, D. R. and Lewis, J. A., "Production,
Formulation and Delivery of Beneficial Microbes for
Biocontrol of Plant Pathogens," Pesticide Formulations
and Application Systems: 11th Volume, ASTM STP 1112,
Loren E. Bode and David G. Chasin, Eds., American Society
for Testing and Materials, Philadelphia, 1992.

ABSTRACT: Industry in the US is currently capable of large-
scale liquid fermentation for production of many common
Microbial pest control agents (MPCAs). Large-scale solid
and semi-solid fermentation methods have also been developed
for MPCAs which do not grow or sporulate well in liquid
fermentation. Recent emphasis in formulation research has
been in the entrapment of MPCA propagules within a cross-
linked matrix of organic polymers such as alginate,
polyacrylamide, or carrageenan. Final formulations are
uniform granules which are compatible with current
agricultural equipment. Most MPCAs survive these
formulation processes well and have a shelf-life of several
months to a few years. These formulations are relatively
inexpensive and permits addition of nutrient bases,
pesticides, or other compounds which may confer a selective
advantage on the MPCA. Other delivery systems include
liquids, dusts, gels, and pastes which are applied to
foliage, soil, wounds, and seeds or seedpieces.

KEYWORDS: biocontrol agents, microbial pest control agents

The goal of production, formulation and delivery of microbial
pest control agents (MPCAs) is to deliver a sufficient number of the
most effective propagules of an MPCA to the best place at the best
time to control a plant pathogen for an acceptable cost. It is
important to note that the goal is not necessarily delivery of the
highest number of propagules. The efficacy of the propagules, and
application timing and placement are often more important than the
population sizes.

Drs. Fravel and Lewis are plant pathologist and soil scientist,
respectively, at USDA, ARS, Biocontrol of Plant Diseases Laboratory,
Beltsville, MD 20705.

There are two major methods of inoculum production: solid (semi-solid) and liquid (deep tank) fermentation.   Solid fermentation is common in laboratory settings.   For example, many fungi and bacteria are commonly grown on petri plates.   Spores, cells or other propagules are then harvested and formulated additionally for biocontrol.   Other types of solid fermentation are also used in biocontrol work.   Grains and straws are often used for production of beneficial microbes [1-2]. This type of fermentation results in a product that is generally used "as is" rather than being formulated further.   These products are likely to obstruct agricultural machinery and thus are probably not feasible for commercial use.   In addition, they require large spaces for processing and storage.   Further, these products must often be applied at extremely high rates in order to achieve control.   Reports of 1%-2% and even 5%-10% amendments are common in the literature.   A 1% amendment on a 15-cm plow layer is 22,400 kg/ha.

Several solid fermentation processes, however, are more commercially acceptable.   For example, a system to produce Sporidesmium sclerotivorum for the control of several pathogens uses technology designed for production of mushroom spawn [3].   Finely ground vermiculite is moistened with liquid medium in an aseptic twin shell blender and macroconidia of Sporidesmium are added.   The twin shell blender rotates to mix the inoculum with the medium.   The mixture is then aseptically bagged.   A paper strip on the plastic bag facilitates gas exchange for the growth of the fungus.   The shelf-life of this formulation is approximately 5 years.

The second major means of producing inoculum is through liquid fermentation.   During the past 30 years liquid fermentation has been developed extensively by industry for the production of antibiotics, enzymes, organic acids and other microbial products [4].   Hence, it appears feasible for industry to produce various beneficial microbes against plant pathogens through liquid fermentation at the present time.   In fact, MPCAs against weeds and insects currently are produced in liquid fermentation [5-6].   Many known biocontrol agents grow well in liquid fermentation, while others do not.   Talaromyces and Sporidesmium, for example, do not sporulate in liquid culture.

For large scale laboratory production, the Biocontrol of Plant Diseases Laboratory, USDA uses carboys containing 15 l of a molasses-based medium with air bubbling up through the medium [7].   The selection of the medium is particularly important.   Industrial waste products that are inexpensive enough to be considered for media include molasses, milk whey, sulfite waste liquor, corn steep liquor, corn sugar, fish meal, bone meal, blood meal, grain, nut and cotton starches and flours [5,8].   A rapid time for the production of effective propagules is generally desirable in order to reduce production costs and decrease chances for contamination.

Once the biomass is formed, it is usually separated from the culture filtrate by filtration, centrifugation or flocculation.   In the Biocontrol of Plant Diseases Laboratory, biomass is separated from the culture filtrates by filtration through cotton dish towels.   The resulting fungal "mats" can be used either wet or dry for further formulation.   Spray or vacuum drying, rather than pan drying, would

probably be preferred by industry because of their efficiency.
However, these two methods reduce viability of the biocontrol fungi
Trichoderma and Gliocladium and pan drying is the preferred method for
these two fungi.

After the inoculum is processed, there are several options for
formulation. In addition to the bulk formulations already discussed,
the options include dusts, liquids, gels and entrapment of propagules
in polymers [9-10]. These formulations may be applied as sprays,
drenches, wound dressings, seed treatments, broadcast and rotovated
into soil, etc. [11-12]. However, these categories are not mutually
exclusive. For example, dusts can be applied as seed treatments and
polymer granules can be rotovated into the soil to alleviate a
particular disease problem. Additional novel technology for
application of biocontrol microbes to seeds is also being developed.
Application of a spore preparation to seed osmotically-primed to
induce rapid germination has resulted in quick plant growth so the
seedling escapes disease [13]. Bio-primed maize seed coated with the
beneficial bacterium Pseudomonas fluorescens was protected from
damping-off in the field [14].

Liquid formulations are versatile and have a number of
applications. Fusarium wilt of greenhouse chrysanthemums was
controlled by an aqueous drench of conidia of Trichoderma [15]. This
MPCA was very effective in preventing reinvasion by the Fusarium
pathogen. Generally, biocontrol is more easily achieved in a
greenhouse setting than under field production since in the greenhouse
biological diversity is lower (thus, less competition) and
environmental fluctuation is minimal. This is not to imply, however,
that liquids have no place in field work. Under production
conditions, eggplants are grown for 6-8 weeks in the greenhouse and
then transplanted into the field with the soilless potting mix in
which they were grown. Good control of Verticillium wilt of eggplant
in the field was obtained by drenching eggplants with Talaromyces
spores before they were transplanted to the field [16].

Aqueous sprays were also used to apply macroconidia of
Sporidesmium to control Sclerotinia of lettuce [17]. Sclerotia of
Sclerotinia are formed on lettuce debris in the field. Sporidesmium
was sprayed onto lettuce plants at the end of the season and then the
debris was plowed under. Even with no additional application of MPCA,
Sporidesmium completely eliminated the pathogen in subsequent crops
even when applied at 0.2 kg/ha. In this example, the key to control
is getting the MPCA to the right place at the right time - that is,
while the sclerotia are still above ground and the pathogen is
accessible. The subsequent aggregate distribution of the sclerotia
then facilitates the spread of the MPCA. As mentioned earlier,
effective MPCA propagules, properly placed, are more important than
sheer numbers of propagules. In each of the last three examples-
chrysanthemum drench, eggplant drench and the lettuce spray-
components of the whole system are considered and the application
method exploits attributes of the MPCA, plant pathogen, and
environment.

Dusts are also used in biocontrol work. Generally, aqueous
suspensions of MPCA propagules are added to a carrier, such as talc or

other types of clay. When dried under a hood overnight, this produces
a fine powder without the need for further grinding. This formulation
has been used to apply Talaromyces and Gliocladium to potato
seedpieces [18-20].

Recently, there has been much interest in the entrapment of MPCA
propagules in polymer prills [21]. The formulation is very flexible
and anything which may confer a selective advantage to the MPCA could
be added such as fertilizers, fungicides, antibiotics, or unusual
nutrients. In general terms, 1%-2% sodium alginate is mixed with
approximately 10% of a carrier and the MPCA propagules in water.
Carrageenan prill are made with 4%-5% carrageenan. Choice of the
carrier depends on the MPCA involved. Trichoderma and Gliocladium
generally require a food base. Wheat bran is commonly used [22].
Other MPCAs, such as Talaromyces, perform better without a foodbase
and pyrophyllite clay is used as the carrier [23]. The polymer, MPCA
and carrier are blended aseptically and added dropwise to a gellant.
For alginate, calcium chloride is generally used as the gellant,
although any di- or trivalent cation is acceptable. For carrageenan,
potassium chloride is used as the gellant. When drops of the alginate
mixture contact the surface of the gellant, the calcium displaces the
sodium causing the formation of a hard shell around the drops. The
prills are then dried to produce a uniform product which may be
incorporated into potting mixes in the greenhouse, or applied
broadcast or in furrow in the field. The prill may be applied with
common agricultural machinery such as fertilizer spreaders.

Organisms differ in their abilities to survive the entrapment
process, and different propagules of the same organism have different
abilities to survive the process. Shelf life of these formulations
generally ranges from a few weeks to several years. Shelf life can
usually be increased greatly by refrigeration. With most fungi in
alginate, we see a rapid decline in viability during the first several
weeks. However, the propagules that do survive this period remain
viable for extended periods [23-24]. For example, Talaromyces will
remain viable for at least 4 years under ambient conditions or
refrigeration. Bacteria typically do not survive well in alginate
because of the gellants used. Thus, carrageenan is preferable for
many bacteria. The prill are not soluble in water and thus act as
point sources in soil. Fungi can grow out of the prill several cm
from these sources and sporulate from the prill. Alginate prill,
formulated with a wide variety of biocontrol fungi by the Biocontrol
of Plant Diseases Laboratory, have effectively reduced plant diseases
in the greenhouse and field [25-28].

Another exciting possibility for the application of beneficial
microbes is through fluid drilling. Fluid drilling has been used for
a number of applications, but there appears to be only one report of
its being used for biocontrol organisms. Polysurf-C gel was used to
apply Trichoderma harzianum and Laetisaria arvalis to reduce diseases
caused by Rhizoctonia solani and Sclerotium rolfsii on vegetable crops
and apple seedlings [29].

There are several formulations of MPCAs currently available
commercially worldwide for plant disease control. However, only
Dagger G, Galltrol, and Norbac are registered for use in this country.

BNAB—T is a mixture of several <u>Trichoderma</u> species used primarily for the control of wood decay. It is formulated both as a dust for wound dressings and as pellets which are inserted into trees [30]. Dagger G contains the bacterium <u>Pseudomonas</u> for control of the seedling disease of cotton. It is formulated as a granule with a peat base [31]. Galltrol and Norbac are powders which are suspended in water into which plants are dipped for control of crown gall [32]. Mycostop is a powder preparation used primarily for treatment of crucifer and cucumber seeds in northern Europe [33]. PG Suspension is a powder which is wetted. In addition, W. R. Grace Co., working with the Biocontrol of Plant Diseases Laboratory has recently registered the first fungus for control of a plant disease. EPA registration was granted for <u>Gliocladium</u> <u>virens</u> to control damping—off of bedding plants caused by <u>Pythium</u> and <u>Rhizoctonia</u>. Quantum 4000 is a <u>Bacillus</u> <u>subtilus</u> preparation that is sold as a growth promotor and not for disease control. Thus, it is not subject to the same regulations as the other products mentioned.

In summary, beneficial microbes are produced through various forms of solid or liquid fermentation. Careful attention to the constituents of the medium and to the incubation conditions results in the maximum quantity of the desirable propagules in the minimum time. The inoculum can be formulated in a number of ways. Some of the most promising appear to be entrapment in polymers, fluid drilling and new technologies with seed treatments. It is clear that this branch of our science is in its infancy. With recent interest in this important area, we should see substantial progress made in the future.

ACKNOWLEDGMENTS

The authors acknowledge with appreciation the help of Martha Hollenbeck in the preparation of this manuscript.

Mention of trademark or proprietary product does not constitute a guarantee or warranty by the US Department of Agriculture, Agricultural Research Service, and does not imply approval over other products that may also be suitable.

REFERENCES

[1]   Cannel, E., and Moo—Young, M., "Solid—state fermentation systems," <u>Process Biochemistry</u>, Vol. 9, 1980, pp. 24—28.

[2]   Carrizales, V. and Jaffe, W., "Solid state fermentation: An appropriate technology for developing countries," <u>Interciencia</u>, Vol. 11, 1986, pp. 9—15.

[3]   Adams, P. B. and Ayres, W. A., "Biological control of Sclerotinia lettuce drop in the field by <u>Sporidesmium</u> <u>sclerotivorum</u>," <u>Phytopathology</u>, Vol. 72, 1982, pp. 485—488.

[4]   Bu'Lock, J. and Kristiansen, B., Eds., <u>Basic Biotechnology</u>, Academic, London, U.K., 1987.

[5]   Bowers, R. C., "Commercialization of microbial biological control agents," in <u>Biological Control of Weeds with Plant</u>

Pathogens, John Wiley & Sons, New York, 1982, pp. 157-173.

[6]   Churchill, B. W., "Mass production of microorganisms for
       biological control," in Biological Control of Weeds with Plant
       Pathogens, John Wiley & Sons, New York, 1982, pp. 139-156.

[7]   Papavizas, G. C., Dunn, M. T., Lewis, J. A., and
       Beagle-Ristaino, J., "Liquid fermentation technology for
       experimental production of biocontrol fungi," Phytopathology,
       Vol. 74, 1984, pp. 1171-1175.

[8]   Zabriskie, D. W., Armiger, W. B., Phillips, D. H., and Alabano,
       P. A., Trader's Guide to Fermentation Media Formulation,
       Memphis, TN.

[9]   Connick, W. J., Jr., Lewis, J. A., and Quimby, P. C., Jr.,
       "Formulation of biocontrol agents for use in plant pathology,"
       in New Directions in Biological Control, Alan R. Liss, New
       York, NY, 1990, pp. 345-372.

[10]   Lewis, J. A., "Formulation and delivery systems of biocontrol
       agents with emphasis on fungi," in The Rhizosphere and Plant
       Growth, Kluwer Academic, 1990, (In press).

[11]   Chet, I., Ed., Innovative Approaches to Plant Disease Control,
       John Wiley & Sons, New York, 1987.

[12]   Papavizas, G. C., "Trichoderma and Gliocladium: Biology,
       ecology, and potential for biocontrol," Annual Review of
       Phytopathology, Vol. 23, 1985, pp. 23-54.

[13]   Harman, G. E. and Taylor, A. G., "Improved seedling performance
       by integration of biological control agents at favorable pH
       levels with solid matrix priming," Phytopathology, Vol. 78,
       1988, pp. 520-525.

[14]   Callan, N. W., Mathre, D. E., and Miller, J. B., " Bio-priming
       seed treatment for biological control of Pythium ultimum
       preemergence damping-off in sh2 sweet corn," Plant Disease, Vol.
       74, 1990, pp. 368-372.

[15]   Locke, J. C., Marois, J. J., and Papavizas, G. C., "Biological
       control of Fusarium wilt of greenhouse-grown chrysanthemums,"
       Plant Disease, Vol. 69, 1985, pp. 167-169.

[16]   Marois, J. J., Johnston, S. A., Dunn, M. T., and Papavizas, G.
       C., "Biological control of Verticillium wilt of eggplant in the
       field," Plant Disease, Vol. 66, 1982, pp. 1166-1168.

[17]   Adams, P. B. and Fravel, D. R., "Economical biological control
       of Sclerotinia lettuce drop by Sporidesmium sclerotivorum,"
       Phytopathology, Vol. 80, (In press).

[18]   Beagle-Ristaino, J. and Papavizas, G. C., "Biological control
       of Rhizoctonia stem canker and black scurf of potato,"
       Phytopathology, Vol. 75, 1985, pp. 560-564.

[19]   Davis, J. R., Fravel, D. R., Marois, J. J., and Sorensen, L. H.,
       "Effect of soil fumigation and seedpiece treatment with
       Talaromyces flavus on wilt incidence and yield, 1983,"
       Biological and Cultural Tests, Vol. 1, 1986, p. 18.

[20]   Keinath, A. P., Fravel, D. R., and Papavizas, G. C., "Evaluation
       of formulations of Talaromyces flavus for biocontrol of
       Verticillium wilt of potato," Biological and Cultural Tests,
       Vol. 5, 1990, p. 30.

[21]   Woodward, J., Immobilized Cells and Enzymes, IRL, Washington,
       D.C., 1985.

[22]   Lewis, J. A. and Papavizas, G. C., "Characteristics of alginate
       pellets formulated with Trichoderma and Gliocladium and their
       effect on the proliferation of the fungi in soil," Plant

Pathology, Vol., 1985, pp. 571–577.

[23] Fravel, D. R., Marois, J. J., Lumsden, R. D., and Connick, W. J., Jr., "Encapsulation of potential biocontrol agents in an alginate–clay matrix," _Phytopathology_, Vol. 75, 1985, pp. 774–777.

[24] Papavizas, G. C., Fravel, D. R., and Lewis, J. A., "Proliferation of _Talaromyces flavus_ in soil and survival in alginate pellets," _Phytopathology_, Vol. 77, 1987, pp. 131–136.

[25] Fravel, D. R., Davis, J. R., and Sorensen, L. H., "Effect of _Talaromyces flavus_ and metham on Verticillium wilt incidence and potato yield, 1984–1985," _Biological and Cultural Tests_, Vol. 1, 1986, p. 17.

[26] Lewis, J. A. and Papavizas, G. C., "Application of _Trichoderma_ and _Gliocladium_ in alginate pellets for control of Rhizoctonia damping–off," _Plant Pathology_, Vol. 36, 1987, pp. 438–446.

[27] Lumsden, R. D. and Locke, J. C., "Biological control of _Pythium ultimum_ and _Rhizoctonia solani_ damping–off with _Gliocladium virens_ in soilless mix," _Phytopathology_, Vol. 79, 1989, pp. 301–366.

[28] Papavizas, G. C. and Lewis, J. A., "Effect of _Gliocladium_ and _Trichoderma_ on damping–off and blight of snapbean caused by _Sclerotium rolfsii_," _Plant Pathology_, Vol. 38, 1989, pp. 277–286.

[29] Conway, K. E., "Use of fluid–drilling gels to deliver biological control agents to soil," _Plant Disease_, Vol. 70, 1986, pp. 835–839.

[30] Ricard, J. L., "Biocontrol of pathogenic fungi in wood and trees, with particular emphasis on the use of _Trichoderma_," _Biocontrol News Information_, Vol. 9, 1988, pp. 133–142.

[31] Currier, T. C., Skwara, J. E., and McIntyre, J. L., "The development of a _Pseudomonas fluorescens_ product (Dagger G) for the control of _Pythium_ and _Rhizoctonia_ on cotton," _Proceedings of the Beltwide Cotton Production Research Conference_, 1988, p. 18–19.

[32] Thompson, J. A., "The use of agrocin–producing bacteria in the biological control of crown gall," in _Innovative Approaches to Plant Disease Control_, John Wiley & Sons, New York, 1987.

[33] Tahvonen, R., "The microbial control of plant diseases with _Streptomyces_ sp.," _EPPO Conference on Strategies and Achievement in Microbial Control of Plant Diseases_, Dijon, France, 1986.

Application Techniques and Pest Control

Daniel R. Wright[1]

THE EXPEDITE® PESTICIDE APPLICATION SYSTEM; A MORE
EFFICIENT, SAFER BACKPACK PESTICIDE APPLICATION SYSTEM.

REFERENCE: Wright, D. R., "The Expedite Pesticide
Application System: A More Efficient, Safer Back-
Pack Pesticide Application System," Pesticide Formu-
lations and Application Systems: 11th Volume, ASTM
STP 1112, Loren E. Bode and David G. Chasin, Eds.,
American Society for Testing and Materials,
Philadelphia, 1992.

Abstract: The Expedite® Pesticide Application System
is a unique backpack application system. The system is
unique in that the packaging, formulation and applicator
are an integrated system. This closed, ready-to-use
system gives a pesticide applicator a way in which to
apply pesticides that reduces exposure to chemicals
through direct handling of the pesticide[1]. The light
weight, accurate delivery, Controlled Droplet
Application (CDA) spray system allows for a more
efficient way to apply pesticides[2] [3].

KEYWORDS: pesticide application, controlled droplet
application, CDA, closed system, pesticide, ready-to-
use, efficiency.

There have been many advancements in the area of pesticide
application during the past years. Many of these advancements
have been in the area of closed system and low volume,
controlled droplet application techniques[2][3][4]. Benefits
from these developments are both greater safety to the
pesticide applicator and economic savings.[5][6] An extension

---

[1] Monsanto Company, 800 N. Lindbergh, St. Louis, MO. 63167.

of this technology has recently been marketed by Monsanto Agricultural jointly with Nomix Corporation in the form of the Expedite® or Nomix® Pesticide Application System. This system was originally developed by Mr. David Gill of Nomix Corporation and jointly refined and marketed in the U. S. and other parts of the world by Monsanto.

The Expedite® system is an integrated pesticide application system. All of the components, sprayer, packaging, and formulation are designed to work together to yield a system that provides the user with a more efficient way in which to apply pesticides. The system consists of a spinning disk sprayer or lance, backpack, and ready-to-use formulated products in a closed packaging system. The pesticide containers fit conveniently into a backpack. There is no direct handling of the pesticide by the applicator required, i.e. no mixing of chemicals.

EQUIPMENT

The Expedite system consists of a rechargeable battery powered sprayer (lance), backpack, battery recharger, calibration cup, two application disks, ready to use pesticide and cleaner. The shaft of the lance is 34 inches in length. This is long enough to allow the applicator to easily reach spots in ornamental beds for spot treatment and to safely distance the applicator from the spray. The lance and backpack are shown in Figure 1.

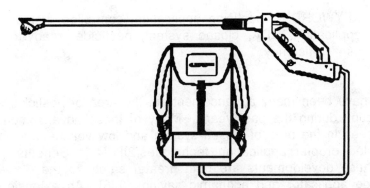

Figure 1. Expedite lance and backpack.

The lance materials of construction, a high impact polymer (Triax®) and aluminum, are designed so as to give a light weight yet rugged piece of equipment. The total weight of the lance is approximately three pounds. It is designed to balance in the hand when held by the handle.

The trigger of the lance is comfortably located in the handle of the lance. There is a trigger lock on top of the handle for continuous spraying.

The control knobs for the lance are easily accessible by their location on the body of the lance. There are two knobs, Figure 2. The top knob controls the spray swath width. The bottom knob controls a pacer/calibration timer that sounds a tone to set a walking speed to assure that the proper amount of pesticide is applied or can act as a timer for calibration. The pacer can be set such that a beep will sound each time a step is to be taken for approximate walking speeds of 1.5, 2.5, and 3 miles per hour. This feature helps insure that the proper amount of pesticide is applied. The calibration feature sounds a beep such that ten beeps equals 30 seconds. This timer is used in conjunction with the calibration cup to accurately calibrate the amount of pesticide to be applied.

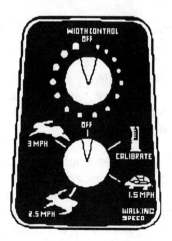

Figure 2. Expedite Lance Control Panel.

The lance contains a rechargeable battery to run the motor and other electrical components. The battery will provide up to 20 hours of use when fully charged. A recharger is provided with the system.

There are two different, interchangeable geometries of disks that can be used. A square disk, for trim and edge uses, that can yield spray swaths of from 6 - 24 inches in diameter and a round, serrated disk that can yield spray swaths from 18 - 36 inches, Figure 3. The spinning disk is rotated by a small motor contained in the spray head of the lance. The speed of rotation is controlled by the swath width knob on the control panel.

Figure 3. Expedite Square and Round Spray Application Disks and Lance Head Cap.

The rate of pesticide delivery is adjusted at the head of the lance, Figure 4. The head cap of the lance rotates to a wide variety of flow rates. A "pointer" on the head indicates what flow setting is being used.

To insure that the proper application rate of pesticide is used, a calibration cup is also provided with the system. The cup snaps snugly onto the head of the lance, Figure 5. It is graduated in units of "multiflows" (a measurement unit unique to the Expedite® system where 1 multiflow = 0.8 mL.)

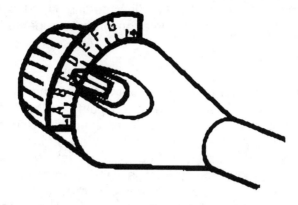

Figure 4. Expedite Lance Head Flow Scale.

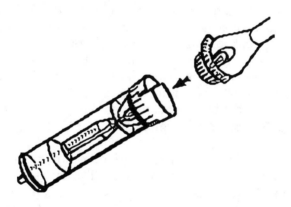

Figure 5. Expedite Calibration Cup and Lance Head.

To calibrate the lance, the applicator can find the proper flow rate to be used for the particular walking speed and spray swath width desired from the product label. The calibration cup is snapped onto the spray head. The pacer knob on the lance control panel is set on the calibrate position, the trigger depressed, and the user counts ten beeps as the pesticide flows into the flow cup. After 10 beeps, 30 seconds, the lance is inverted with the flow cup upright and the amount of

material is easily read. The trigger is then depressed and pesticide drained back into the lance. There is no waste of pesticide from calibration. If the calibration (flow rate) is correct the user is ready to apply the pesticide. If the flow is too great or too small, the flow setting is adjusted by rotating the head cap and the calibration process repeated.

PACKAGING

The pesticide formulation is delivered to the spinning disk through gravity flow. There is no pump in the system. This required that a new concept in packaging of the pesticides be developed. As the package is emptied during application it would be necessary to either allow air into the package or have the package collapse to keep a vacuum from forming. Any type of air vent in the package would introduce the possibility of the pesticide leaking from the vent. Even if a one-way valve were used there is the possibility of either leaking and exposure of the applicator to the pesticide or becoming clogged and causing an interruption of flow. A "bag-in-a-box" package was chosen as an alternative type of package to give uninterrupted flow of the pesticide to the lance and minimize the possibility of exposure of the applicator to the pesticide through leakage from the container. The container is composed of a box outer shell around an inner bag. The bag collapses as the product flows from the container, keeping a vacuum from building as the container empties.

The container is connected to the lance through a specially designed bung, as shown in Figure 6. The bung is pierced by the fitting on the lance hose. There is a C-clip that is secured after connection of the lance to the container to keep the lance attachment from being pulled from the bung. The bung can be resealed with an attached cap if only a portion of the product is used.

Figure 6.　Expedite Lance to Container Connection.

The volume of each container is five liters. This is a sufficient amount of chemical to treat from 2/3 - 1 1/3 acres depending on the product and treatment rate.

A container of the pesticide fits into a backpack. The backpack is constructed of water resistant fabric with an aluminum support frame, much like a hiking backpack. Padded shoulder straps and a waist strap are adjustable for a comfortable fit. The backpack and chemical weighs approximately 12 lbs. When this is compared to the approximately 25 - 30 lbs weight of a full conventional 3 gallon backpack sprayer, the reduced worker fatigue advantage of the Expedite system becomes very apparent.

FORMULATIONS

The Expedite pesticide formulations are specifically designed for use with the Expedite application equipment. They are ready-to-use, meaning that there is no mixing of chemicals required by the applicator. The flow rate of the product need only be set to match the application rate, spray swath width and walking speed of the applicator.

The products are formulated such that the rheological properties of the formulation yield the best possible flow properties and spray pattern with a gravity fed, spinning disk atomizer. The spray pattern can also be described as the best possible coverage by the pesticide within the target area. This can be translated into the efficacy of the product, especially when the particular pesticide being applied is a contact herbicide such as glyphosate. The spray pattern is (as with all spinning disk atomizers) a circular, "donut" shaped pattern.

In order to obtain a "controlled" spray droplet formation, the spray drops must form from ligaments extending from the spinning disk atomizer[11]. The formation of ligaments from the spinning disk is controlled by the rheological and surface properties of the formulation as well as the interaction of the formulation with the surface of the spinning disk and it's angular velocity. When ligaments are formed, the spray pattern consists of fairly uniform size spray droplets and the spray pattern will be visible as a "donut" pattern with well defined outer and inner edges. If the spray drops are formed either directly from the disk or from sheets of liquid flooding from the disk, the spray pattern will be very diverse with a wide range of droplet sizes and a widely spread pattern consisting of very large to extremely fine drops. With the larger size droplets the likelihood of poor target area coverage (fewer drops per spray area), and hence poor efficacy, is probable. Very fine spray particles can lead to spray drift to non-target areas.

To enhance the accuracy of application the products are formulated so that the spray is visible to the applicator. This is accomplished by formulating the products such that they are an opaque color, preferably white. It has been found through much testing and observation of sprays that a white color is most visible against foliage. A clear spray drop, even if colored, is virtually invisible while in flight. Visibility of the spray is most important when the application is made at a smaller spray width, such as in trim and edge uses, where it is important to have a narrow, accurate delivery of the pesticide. This allows the applicator to know if, when and where he/she is applying the pesticide. This makes the application of the pesticide more accurate because the applicator can tell where the edges of the spray pattern are and reduces the possibility of spraying a non-target area.

There are two products currently sold commercially. Expedite Grass and Weed Herbicide is a glyphosate containing post-emergence, herbicide and Expedite Broadleaf Herbicide is a combination of 2,4-D and MCPP isooctylesters for broadleaf weed control in turf. Other formulations under development are various types of herbicides, insecticide, and fungicide products.

BENEFITS

The benefits of the Expedite pesticide application system are both economy and safety to the user. The economic benefits come from more efficient use of the applicator's time as well as more efficient use of the pesticide.

No time is required for mixing chemicals. Each ready-to-use container of the low volume Expedite application covers a larger area than one full conventional backpack. This reduces the amount of time lost due to returning to a water source to refill a conventional sprayer with water and remix chemicals in the coarse of a treatment . The low volume application of the Expedite system can treat as much area with one container as 15 - 25 refills of a conventional backpack.

A more accurate and economic use of pesticides can result from the system because of the flexibility and CDA technology of the application.[5] The user can match the application rate to their desired walking speed and spray swath width. Since the spray is visible, the likelihood of off-target injury due to error is reduced because the applicator knows where the spray is going. Additionally, the CDA spray reduces the possibility of drift [2][5][7][8][9]. The ready-to-use pesticide formulations reduce the likelihood of mixing errors which could result in over-application and waste of the pesticides. The pacer feature of the lance also aids in reducing the likelihood of over-application by matching the applicator to a walking speed that insures a correct product application rate.

Safety benefits from the Expedite system result from reducing the handling of chemicals by the applicator. The formulations are ready-to-use. No mixing of chemicals is required. This greatly reduces the risk of chemical exposure to the applicator[1][6][10]. Additional safety benefits are from the CDA atomization used in the system. This type of atomization system produces fewer fine droplets (<50 - 100 microns) in the spray that can drift to the applicator[2]. The light weight of the system greatly reduces worker fatigue and possibility of injury.

The Expedite Pesticide Application System offers many benefits to the pesticide applicator. The system is an integrated system in which all of the components work together to give the user the benefits of a closed, ready to use packaging system with lightweight CDA technology. These benefits are safety to the user as well as economic and accurate use of pesticides. The packaging, formulations and equipment yield one of the most advanced systems developed to date for backpack pesticide application.

REFERENCES

[1]   Grover, R.; Cessna, A. J.; Muir, N. I.; Ridel, D.; Franklin, C. A.; "Pattern of Dermal Deposition Resulting From Mixing/Loading and Ground Application of 2,4-D dimethylamine salts", ASTM Spec. Tech. Publ. ASTTA, 8, 1988, 989, pp. 625 - 9.

[2]   Bals, T. E.; "Economical Pesticide Application: The Reason for Controlled Droplet Application", Pesticide Formulation and Application Systems: Seventh Volume, ASTM STP 968, G. B. Beestman and D. I. B. Vander Hooven, Eds., American Society for Testing and Materials, Philadelphia, 1987, pp. 133 - 138.

[3]   Merry, J. C.; "The Effects of Droplet Size and Application Method on the Activity of Pre-emergence Herbicides"; Imperial College of Sci. Tech., Univ. London, London, UK, 1987, 36(1); pp. 270-271.

[4]   Schmidt, E.; "The BIRKY, A New Concept for Better Spraying"; Proceedings of the Second Biannual Conference, West African Weed Science Society, M. Deat,; P. Marnotte, Eds., West African Weed Science Society, 1983, pp. 377 - 382.

[5]   Bals, E. J.; "The Reasons for CDA (Controlled Droplet Application)", Proceedings 1978 British Crop Protection Conference - Weeds, 1978, pp. 659 - 666.

[6]   T. L. Lavy, J. D. Walstad, R. R. Flynn, J. D. Mattice; J. Agric. Food Chem.; 1982, 30, pp. 375 - 381.

[7]   Bals, E. J.; "Where Have All the Droplets Gone?"; Proceedings of the Seventh Australian Weeds Conference, 1984, Vol. II, pp. 81 - 85.

[8]   Wiltse, M. G.; "Developments in Controlled droplet Application for Weed Control", Abstracts, 1983 Meeting of the Weed Science Society of America, 1983, pp. 53.

[9]   Scoresby, J. R.; Nalewaja, J. D.; "Drift With the Controlled Droplet Applicator"; Proceedings North Central Weed Control Conference, 1982, (Vol. 36), pp. 8 - 9.

[10]   Matthews, G. A.; "Protective Clothing and Machinery"; Chemistry and Industry, 1985, 18, pp. 617 - 619.

[11]   Masters,K; Spray Drying Handbook, Halsted Press, 1972, pp 175 - 214.

Andrew C. Chapple[1], Roger A. Downer, and Franklin R. Hall

EFFECTS OF PUMP TYPE ON ATOMIZATION OF SPRAY FORMULATIONS

REFERENCE: Chapple, A. C., Downer, R. A., and Hall, F. R., "Effects of Pump Type on Atomization of Spray Formulations," Pesticide Formulations and Application Systems: 11th Volume, ASTM STP 1112, Loren E. Bode and David G. Chasin, Eds., American Society for Testing and Materials, Philadelphia, 1992.

ABSTRACT: The effects on the atomization process with respect to droplet spectra were investigated for three pump systems (a rotary vane pump, a piston pump, and a diaphragm pump) and compared with an air pressure system. The test substances evaluated were water, Ortho X-77 (0.0625% and 0.5%), Windfall and Nalcotrol polymers, and two placebo (blank) insecticides (an emulsifiable concentrate and a wettable powder). All formulations were tested at field rates. All work was done using a TeeJet XR8004VS flat fan nozzle, and a Hardi 1553-12 hollow cone nozzle with a grey swirl. The effect of recirculation of the formulations on the eventual droplet spectra was also investigated. All other spraying parameters were kept constant. Droplet spectra were assessed along the long axis of the spray pattern using an Aerometrics PDPA-100 system, and a Velmex "Unislide" xyz-positioner.

Results indicate that droplet spectra of formulations containing polymers shift significantly with time of recirculation: $D_{v0.5}$ decreased by as much as 300 μm for Nalcotrol. The time taken for the $D_{v0.5}$ to stabilize depended upon pump type. Water, Ortho X-77, and the two blank formulations remained unaffected by pump type or recirculation. The effect of different mixing regimes prior to atomization was also investigated for Windfall and Nalcotrol.

KEYWORDS: Droplet atomization, lasers, pumps, adjuvants, drift, nozzles, drop spectra.

[1] A.C. Chapple, Graduate Research Associate; R.A. Downer, Post-Doctoral Research Associate; and F.R. Hall, Head and Professor; Laboratory for Pest Control Application Technology, The Ohio State University, OARDC, Wooster, OHIO 44691-4096

INTRODUCTION

In recent years, there has been an increase in the attention paid to pesticide targeting and, in particular, the mis-targeting of pesticides, whether as drift or loss to groundwater. One facet of agricultural spraying that is unlikely to change in the next decade is the reliance upon hydraulic systems - flat fan and hollow cone nozzles - as the mainstay of pesticide application. One approach to controlling off-target losses as drift is to alter the droplet spectrum of a given nozzle to produce less fine or "driftable" droplets. In this work, this is defined as drops < 100 μm in diameter. There are a number of drift control additives on the market (eg. Nalcotrol); however, it should be noted that polymers are not limited to drift additives, but are used in agricultural formulations for a variety of purposes [1,2].

Preliminary data gathered in 1989 using Windfall (a linear alkyl epoxide polymer) and Nalcotrol (a polyvinyl polymer) indicated that by changing from a rotary vane pump to a piston pump and altering the temperature of the water used for dilution, distinct differences in droplet spectra could be obtained, especially with Nalcotrol. It was suspected that both changes had an effect on droplet spectra, and that the difference in shear stresses between the rotary vane pump and the piston pump was responsible for the majority of the effects observed. An alternative hypothesis to explain the observed changes was either incomplete mixing of the adjuvants, or stratification of the polymer in the spray volume.   The reduction in the $D_{V0.5}$ for Nalcotrol after recycling through a pump has been observed by Bouse, Carlton, and Jank [3], using a D6-46 hollow cone in a high speed airstream to simulate aerial spraying, and a gear pump as the delivery system. They observed differences in $D_{V0.5}$ of up to 120 um, whereas using ground-rig equipment and lower pressures, we observed much larger decreases in $D_{V0.5}$, on occasion > 500 um.   These observations and the increased demand on the facilities   at   LPCAT   to   undertake   studies   of   polymer-based formulations/spray adjuvants suggested a reassessment of the methodology being used for droplet spectra assessments.

A number of questions require answering if formulations are to be assessed   consistently.   Was   this   phenomena   limited   to   polymeric adjuvants? Were both polymers equally sensitive? As the use of polymers increases in formulations and as registration authorities require more data concerning the spraying characteristics of pesticide formulations, it has now become essential to determine the impact of variables other than spray height, pressure, and nozzle type on the eventual droplet spectra.

Our objectives in this research were to determine:

1.   The effect of pump type (rotary vane, piston, and diaphragm) on   atomization   of   a   selection   of   spray   adjuvants   and   blank formulations;
2.   Whether thorough mixing of the formulation/adjuvant prior to spraying had any effect on the eventual droplet spectra.

MATERIALS AND METHODS

Test Substances:

   Three adjuvants and two blank formulations were evaluated. All
dilutions of adjuvants and formulations were done on a % v/v basis. The
three spray adjuvants were: Ortho X-77 (Chevron, Richmond CA) at 0.0625%
and 0.5%; Windfall (Terra International, Waco, TX) at 0.375%; and
Nalcotrol (Nalco Chem. Co., Naperville, IL) at 0.0625%. The two
formulations were blanks of commonly used insecticides: an emulsifiable
concentrate, Pydrin 2.4 EC (E.I. DuPont De NeMours, Wilmington, DE) at
0.4%; and a wettable powder, Ambush 25 WP (ICI Americas, Wilmington, DE)
at 0.225%. The concentrations reflect normal field rates. As only a
small quantity of the Ambush blank was available, this formulation was
not run through the diaphragm pump because large (> 60 L) quantities of
spray mixture were required for the pump to operate realistically. Tap
water was used as a standard and for initial calibration of the droplet
sizing equipment.

Nozzles:

   Two nozzles were used: a TeeJet XR8004VS extended range flat fan
(Spraying Systems, Wheaton, IL), and a hollow cone 1553-12 with grey
swirl plate (Hartvig Jensen & Co.$^A/_s$, Ontario, Canada). The former was
used at 276 kPa (30 psi), the latter at 520 kPa (75 psi) and 1034 kPa
(150 psi). Pressure was measured using Ashcroft gauges (0-414 kPa,
± 1.7 kPa and 1379 kPa, ± 17 kPa). Flow rates were checked with water
and found comparable to manufacturers' recommendations. Strainers
(50 mesh) were used with all nozzles. A single experiment was done using
a TeeJet D6-45 nozzle at 1034 kPa and Nalcotrol.

Pump Systems:

   Three pump systems were used, with a 10 gallon (37.9 L) Cornelius
pressure can as a standard. The container had a maximum pressure rating
of 896 kPa (130 psi), and hence for the rotary vane and piston pumps, the
hollow cone measurements were taken at two pressures - 517 and 1034 kPa -
to allow true nozzle comparisons. The three pumps used were a piston
pump (Cat) with a throughput of 9.4 L/min; a rotary vane pump (Procon)
with a throughput of 8.9 L/min; and a diaphragm pump (Hypro) with a
throughput of 46.2 L/min. The throughputs were measured as the
recirculating volume when not spraying. The rotary vane and piston pumps
were laboratory equipment and of necessarily low output; the diaphragm
pump was used directly from a 500 L research orchard sprayer. All pump
systems had simple recirculating systems, and the rate of recycling was
determined prior to the experiment using water.

   The spray volume was made up according to manufacturer's
instructions. Temperature of the spray mixtures was measured at various
times during the experiment, and even after multiple recycling through
the pumps, did not alter by more than ±3° C over any experiment (with one
exception of 5° C). All lines and containers were rinsed thoroughly
between formulation runs, using hot and cold water.

For those test substances shown to be sensitive to shear stress, experiments were done to determine if mixing rather than the shearing effect of the pumps could explain the differences observed. All mixing was done in the pressure can or similar container. Six different methods were used to mix the formulations:

- previous laboratory practice in dealing with small volumes of Nalcotrol - ie. diluting the Nalcotrol in a stream of water, but at a higher concentration than required, then diluting further;
- as per manufacturers instructions (Nalcotrol added as a thin stream to the in-flowing water) as would occur in the field;
- made up as per manufacturer's instructions for 15 L, and shaken vigorously by hand in the pressure can;
- made up as per manufacturer's instructions and stirred for 30 minutes with an air-powered paddle stirrer;
- made up as per manufacturer's instructions and passed through the rotary vane pump, bypassing the recirculation; and
- made up as per manufacturer's instructions and sprayed out using the rotary vane pump after recycling.

Droplet Spectra:

All measurements of droplet spectra were made using an Aerometrics (SunnyVale, CA) phase doppler laser velocimeter (PDPA-100), measuring across the diameter of the hollow cone nozzle and the long (or x-) axis of the flat fan nozzle. The latter has been shown to give an acceptable assessment of the full spray pattern for a flat fan nozzle. Spray height was 30 cm. Prior to the experiment, the PDPA-100 was checked for accuracy using a near-monodisperse droplet generator [4] and comparing the PDPA-100 results with water sensitive paper (Ciba-Geigy, Basle, Switzerland) and magnesium oxide slides [5]. (If the PDPA-100 is found not to agree with water sensitive papers or MgO slides, then it is returned to Aerometrics for a factory calibration.) Droplet measurements were found to be within manufacturer's tolerances ($\pm$ 2% for $D_{v0.5}$), for the limits of the calibration technique. The PDPA-100 was utilized with the detector voltage set at 325 volts, and a velocity offset of 20 m/s. Diameter range was set by trial and error when droplets greater than 700 µm were measured. The top limit was set such that the number of small droplets falling outside the measurement range was minimized, even where this meant failing to measure oversize drops. For example, 1300 µm was set as the maximum drop diameter for the Nalcotrol sprays, even though this meant, on one occasion, discarding up to 13% of the spray as oversize. It should be stressed that discarding large droplets from the measurement of the droplet spectra underestimates the droplet spectra distribution, but ensures that the lower end of the range including the 'driftable component' was measured accurately. This was deemed to be an acceptable compromise, as errors in measurement would therefore underestimate $D_{v0.5}$, giving conservative estimates of change.

Validations (ie. acceptable signals) were maintained at better than 92% throughout the entire experiment, with the only exceptions being some runs with Nalcotrol. Where validations fell below 90%, the cause was determined and this was always due to excessive numbers of drops out of

The flat fan nozzle was mounted over the probe volume on an xyz-positioner ("Unislide", Velmex Inc., E. Bloomfield NY), moved out of range of the detector, and traversed through the probe volume on the long axis of the spray cloud. Unlike the hollow cone, the flat fan could be removed and replaced accurately. The hollow cone was mounted once, and was not moved from the mounting for the duration of all the work done using this nozzle. Irregularities in the structure of the spray cloud from a hollow cone, easily seen by eye, were particularly noticeable for the D6-45 using Nalcotrol. The speed of the traverse was set to ensure sufficiently large numbers of measurements per pass (from previous experience, > 10,000). Where this was not possible, multiple passes were made, and the data merged using the PDPA-100 software.

Day-to-day alignment of the components of the measuring system was done by spraying tap water using the pressure can (276 kPa flat fan, 517 kPa hollow cone) and adjusting either the traverse co-ordinates of the nozzle or the focus of the PDPA-100 receiver. Alignment was considered acceptable when results ($D_{V0.5}$, $D_{N0.5}$, % validations) were obtained within an acceptable tolerance for the x-axis traverse of the spray cloud as compared with the previous and original 'setup' runs. Over the period of the experiments, the setup varied by ± 3.4 μm / ± 3.9 μm for $D_{V0.5}$ / $D_{N0.5}$ for the flat fan nozzle (99% CI), and not more than ± 5.1 μm / 2.9 μm for the hollow cone nozzle.

All statistical analysis was done using Cohort Software (Berkeley, CA), using ANOVA and SNK (Student Newman Keul's test) for separation of means.

RESULTS

Effects on Test Substances:

The $D_{V0.5}$ and $D_{N0.5}$ for all formulations, adjuvants, and water sprayed using the pressure can and the XR8004VS nozzle are given in Table 1. The $D_{V0.5}$ showed considerable variation between formulations, but with the exception of Pydrin EC blank and Nalcotrol, the $D_{N0.5}$'s for the flat fan data did not differ greatly from that of water. At 517 kPa through the hollow cone nozzle, Windfall also had a higher $D_{N0.5}$ than the general trend.

Results for formulations sprayed through the hollow cone are shown in Table 2. Droplet spectra were measured for the formulations immediately after mixing and after recycling for 30 minutes.

Results for the effect of pump type on formulation as sprayed through the XR8004VS nozzle at 276 kPa are presented in Figures 1 through 3, for Ortho X-77 at 0.0625%, Windfall, and Nalcotrol, respectively. Only two measurements were taken for the diaphragm pump, at 0 and 30 mins. These are represented on the graph by "D". Ortho X-77 was unaffected by pump type. Windfall showed a small decline in $D_{V0.5}$ with the rotary vane pump, whereas the piston and diaphragm pump lowered the $D_{V0.5}$ immediately. Nalcotrol showed a large decrease in $D_{V0.5}$ with timeof recirculation through all of the pumps but not when left standing in the pressure can for an equivalent time. Ortho X-77 at 0.5%, Pydrin EC blank, and Ambush WP blank behaved exactly as Ortho X-77 at 0.0625%, with the difference that their $D_{V0.5}$ and $D_{N0.5}$ measurements differed (Table 1).

TABLE 1 -- $D_{V0.5}$ and $D_{N0.5}$ (μm) for seven formulations, sprayed through a pressure can with an XR8004VS nozzle at 276 kPa, and a 1553-12 hollow cone nozzle with grey swirl at 517 kPa.

| | | NOZZLE: | | | |
| | | XR8004VS | | 1553-12 (Grey) | |
	Conc.	$D_{V0.5}$:	$D_{N0.5}$:	$D_{V0.5}$:	$D_{N0.5}$:
Water	-	303.4	111.2	186.7	72.5
Ortho X-77	0.0625%	322.0	118.5	192.4	78.5
Ortho X-77	0.5%	281.6	104.4	171.3	70.3
Pydrin EC(blank)	0.4%	334.2	167.6	210.8	107.7
Ambush WP(blank)	0.225%	305.2	110.7	191.7	77.3
Windfall	0.375%	444.4	113.7	364.6	98.6
Nalcotrol	0.0625%	514.5	134.2	390.0	103.2

The $D_{V0.5}$ for the test substances sprayed through the flat fan nozzle are shown in Figures 4 and 5 for the piston and rotary vane pumps respectively. (Data points on the line for Nalcotrol indicate number of recirculations through pump.) Nalcotrol exhibits an initial $D_{V0.5}$ far higher than the other formulations, but this declines with recycling. Windfall shows a similar but lesser response to recycling through the piston pump, comparable with Nalcotrol.

In the rotary vane pump, the declining $D_{V0.5}$ for Nalcotrol levels out at the general range of $D_{V0.5}$'s for all the formulations. In the piston pump, Nalcotrol $D_{V0.5}$ was still decreasing after 44 minutes recycling, not reaching a plateau before all the liquid had been sprayed. Through the rotary vane pump, the remaining test substances show a flat response to recycling time.

Effects on Formulations:

The data for the mixing experiments are shown in Tables 3, 4, and 5. Table 3 shows the effect of different mixing regimes on the $D_{V0.5}$ and $D_{N0.5}$ for Nalcotrol. $D_{V0.5}$ and $D_{N0.5}$ are significantly lowered by spraying through the rotary vane pump. Table 4 shows the effects on $D_{V0.5}$ and $D_{N0.5}$ for Nalcotrol when un-stirred, stirred for 30, 45, and 60 min, and left standing for 60 min. There was a significant shift in $D_{V0.5}$, comparing un-stirred, left standing for 60 min, and stirred for 60 min. $D_{N0.5}$ was increased by either stirring or standing. Table 5 shows data for the effect on $D_{V0.5}$ and $D_{N0.5}$ of stirring, the action of the rotary vane pump alone (recycle bypassed), and the action of the pump recycling on Nalcotrol and Windfall. The rotary vane pump significantly reduced $D_{V0.5}$ for both Nalcotrol and Windfall, and reduced the $D_{N0.5}$ for Nalcotrol. Mixing only affected the $D_{N0.5}$ for Nalcotrol.

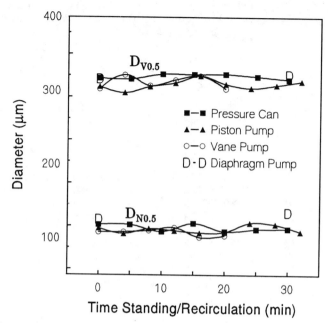

Figure 1.   The effect of pump type and recirculation on the $D_{V0.5}$ and $D_{N0.5}$ for Ortho X-77, at 0.0625% concentration, an XR8004VS nozzle, at 276 kPa.

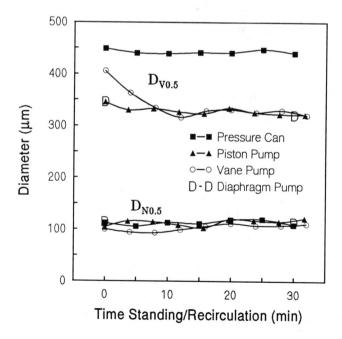

Figure 2.   The effect of pump type and recirculation on the $D_{V0.5}$ and $D_{N0.5}$ for Windfall, at 0.375% concentration, an XR8004VS nozzle, at 276 kPa.

TABLE 2 -- $D_{V0.5}$ and $D_{N0.5}$ for six formulations sprayed with a 1553-12 hollow cone at 517 and 1034 kPa, with recycling through pump.

a. Hollow cone at 517 kPa

Test Substance	Recycling Time(min)	Pressure Can $D_{V0.5}$	$D_{N0.5}$	Piston Pump $D_{V0.5}$	$D_{N0.5}$	Diaphragm Pump $D_{V0.5}$	$D_{N0.5}$
Ortho X-77	0	194.8	80.0	205.0	76.6	207.7	88.3
0.0625%	30	188.9	77.1	203.0	78.2	216.8	104.0
Ortho X-77	0	174.3	72.3	178.8	79.2	177.7	79.8
0.5%	30	168.0	69.0	178.4	79.3	175.8	80.9
Pydrin EC(blank)	0	209.5	106.7	223.9	126.5	225.9	128.9
0.4%	30	218.5	121.0	222.2	128.4	224.9	129.8
Ambush WP(blank)	0	190.7	78.7	189.2	85.6	...	...
0.225%	30	192.6	76.8	190.2	85.5	...	...
Windfall	0	369.4	95.2	221.2	83.8	215.6	82.7
0.375%	30	360.2	102.4	193.9	80.8	192.5	78.6
Nalcotrol	0	384.8	102.8	361.0	98.9	520.7	125.9
0.0625%	30	391.7	98.7	201.5	79.6	203.5	83.9

b. Hollow cone at 1034 kPa

Test Substance	Recycling Time(min):	Piston Pump $D_{V0.5}$	$D_{N0.5}$	Vane Pump $D_{V0.5}$	$D_{N0.5}$	Diaphragm Pump $D_{V0.5}$	$D_{N0.5}$
Ortho X-77	0	165.5	78.8	160	90.4	171.8	92.3
0.0625%	30	166.8	82.8	...	...	167.4	91.3
Ortho X-77	0	160.5	86.8	152.4	86.5	161.1	89.9
0.5%	30	159.3	85.2	...	...	158.3	85.6
Pydrin EC(blank)	0	168.4	101.8	166.6	101.2	168.8	100.0
0.4%	30	172.1	101.3	...	...	171.5	103.8
Ambush WP(blank)	0	168.6	90.9	164.3	88.8	...	...
0.225%	30	167.7	90.3	...	...	...	...
Windfall	0	164.7	78.5	186.4	83.1	166.0	78.4
0.375%	30	164.3	85.8	...	...	162.0	83.8
Nalcotrol	0	288.8	80.0	348.2	96.7	248.2	94.3
0.0625%	30	180.5	74.2	...	...	169.2	89.9

TABLE 3 -- $D_{V0.5}$ and $D_{N0.5}$ for Nalcotrol for different mixing regimes, using an XR8004VS nozzle at 276 kPa, sprayed with a pressure can.

Mixing Regime	$D_{V0.5}$		$D_{N0.5}$	
Laboratory Mix	524.9	b [t]	124.1	a
50 Inversions	546.5	a	127.2	a
Recycled and sprayed through vane pump.	425.5	c	113.0	b

[t] : Letters not different in the same column indicate no significant difference at P = 0.05.

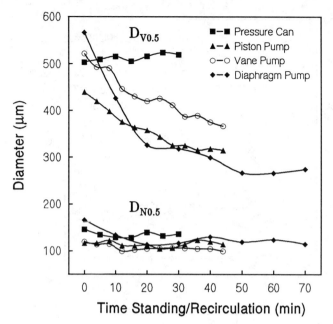

Figure 3.   The effect of pump type and recirculation on the $D_{V0.5}$ and $D_{N0.5}$ for Nalcotrol, at 0.0625% concentration, an XR8004VS nozzle, at 276 kPa.

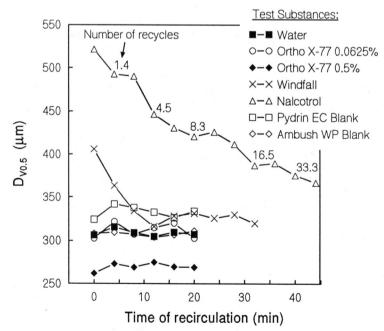

Figure 4.   The effect on the $D_{V0.5}$ and $D_{N0.5}$ for all formulations sprayed through a piston pump, an XR8004VS nozzle, at 276 kPa.

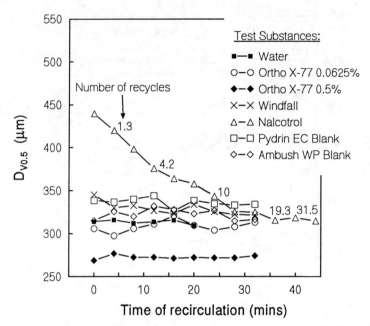

Figure 5.   The effect on the $D_{V0.5}$ and $D_{N0.5}$ for all formulations sprayed through a rotary vane pump, an XR8004VS nozzle, at 276 kPa.

Table 6 shows the $D_{V0.5}$, $D_{N0.5}$ and recirculations for Nalcotrol sprayed through a TeeJet D6-46 hollow cone nozzle at 1034 kPa. In keeping with the manufacturer's instructions, the concentration of Nalcotrol was raised to 0.125%. The results show a decline in $D_{V0.5}$, reaching a plateau at approx. 25 minutes, by which time, the liquid has been sprayed has been through the pump approx. 8.3 times.

TABLE 4 --   Effect of period of stirring on $D_{V0.5}$ / $D_{N0.5}$ for Nalcotrol using an XR8004VS nozzle at 276 kPa, sprayed with a pressure can.

	$D_{V0.5}$	$D_{N0.5}$
Un-stirred	608.4 c †	118.5 b
30 min stirring	679.5 ab	129.6 a
45 min stirring	672.0 ab	127.7 a
60 min stirring	688.6 a	133.6 a
60 min standing	645.8 b	130.0 a

† : Letters not different in the same column indicate no significant difference at P = 0.05.

TABLE 5 --   Effect of stirring or recycling on $D_{V0.5}$ / $D_{N0.5}$ for Nalcotrol and Windfall, sprayed using an XR8004VS nozzle at 276 kPa with a pressure can or vane pump.

	Nalcotrol:		Windfall:	
	$D_{V0.5}$	$D_{N0.5}$	$D_{V0.5}$	$D_{N0.5}$
Manufacturers' Recommendations:	749.4 b [†]	107.4 a	468.1 a	107.0 ns
30 minutes mixing:	825.5 a	110.7 a	461.6 a	112.4 ns
Vane pump, no recycling:	511.4 c	88.6 b	361.9 b	110.0 ns
Vane pump, 20 min recycling:	529.2 c	87.8 b	373.2 b	106.6 ns

[†] : Letters not different in the same column indicate no significant difference at $P = 0.05$. (ns indicates not significant.)

TABLE 6 --   Effect on $D_{V0.5}$ and $D_{N0.5}$ for Nalcotrol (0.125%) sprayed through D6-46 nozzle, at 1034 kPa, with a diaphragm pump (from an orchard sprayer).

Time	0	5	10	15	20	25	30	35
Recycled	0	1.2	2.5	4.1	6.0	8.3	11.4	15.9
$D_{V0.5}$	533.0	316.4	259.6	237.1	226.6	223.5	216.2	222.1
$D_{N0.5}$	140.8	130.3	127.4	124.2	124.6	124.9	126.5	128.3
% Volume[†] < 114 μm	1.3	3.8	5.5	6.7	7.3	7.4	7.5	6.8

[†] as computed from the PDPA-100 program.
Water: $D_{V0.5}$ 216.4 μm; $D_{N0.5}$ 121.7 μm, % Volume[†] < 114 μm 7.3%

DISCUSSION

The results show that the four pumps did not affect the majority of the formulations tested (Ortho X-77, Pydrin EC blank, Ambush WP blank). Nalcotrol and, to a lesser degree, Windfall were both affected by pump action. Terra Int. suggest in their product information for Windfall that "one or two cycles through the tank is sufficient for complete blend", and this would appear to be the case. However, from Table 5 it can be seen that it is the passage through the pump that causes the reduction in $D_{V0.5}$ / $D_{N0.5}$, not the mixing process. In point of fact, for Nalcotrol, the $D_{V0.5}$ / $D_{N0.5}$ is _increased_ by mixing. Pump type is unimportant as a variable in determining $D_{V0.5}$ / $D_{N0.5}$: the critical factor being the shear stresses involved in any pumping operation. On the basis of this study, the critical factor (or variable) affecting $D_{V0.5}$ is shear stress.

Nalcotrol gave variable results over the series of recycling experiments. On average, $D_{V0.5}$ would start at approx. 550 µm and fall to approx. 350 µm. On one run through the piston pump, $D_{V0.5}$ started at 858.0 and fell to 293.2 µm (hollow cone, 517 kPa), the recycling only occurring during droplet measurement. The higher starting $D_{V0.5}$ was due to the higher concentration of Nalcotrol used (0.125%), as per manufacturer's recommendations for spraying using hollow cones at 345 to 1034 kPa (50 to 150 psi).

In terms of how many times the liquid will pass through the pump prior to being sprayed, recycling in a tank follows an exponential curve. At the beginning of a spray run, very little of the tank contents is moved by the pump, but by the end of the spray session, the liquid remaining will have passed through the pump many times. For example, a 20 nozzle boom, output 1.5 L/min per nozzle, using a 2000 L tank and a pump capable of 100 L/min throughput will recycle 70 L/min, while spraying 30 L/min. The final 30 liters of spray liquid will have passed through the pump 12.8 times. The critical factor is the proportion of the liquid of the pump's output that is recycled. If this is reduced to 50 L/min total output, the last 50 liters of spray liquid will have passed through the pump only 3.6 times. In this respect, the experiments exaggerate the number of times a spray liquid is recycled, yet in the case of Nalcotrol, Table 6 shows that by 4.1 cycles through the pump, $D_{V0.5}$ and $D_{N0.5}$ have fallen to almost that of water, and by 11.4 cycles, cannot be distinguished from water.

Preliminary work with a mixture of Pydrin EC blank and Nalcotrol suggests that the Nalcotrol overrides the effects of the Pydrin EC blank, increasing $D_{V0.5}$ and decreasing $D_{N0.5}$. Driftable component for the mixture is reduced by approx. 30%. to 0.87%. In the process, the % volume above 350 µm is raised from 54% to 93.5%, with 35% of the volume contained in droplets > 1000 µm in diameter. Whether a 1 mm diameter droplet has a biological value in terms of spraying is debatable.

The work raises several issues concerning the assessment of formulations for their droplet spectra. Clearly, many pesticide formulations and adjuvants will be largely unaffected by the pressurizing system. However, polymers are being used for a variety of purposes in formulations, in addition to drift control, and an assessment of the impact of such polymers on the droplet spectra of a spray produced by a given nozzle should represent the eventual use of that formulation in the field. Therefore, we would suggest that any formulation that is suspected of being sensitive to shear stress should be recycled through a pump system (of any design) at least 20 times prior to droplet spectra measurements, and that these be compared with the formulation sprayed through a pressure can. Lacking this information, misleading conclusions regarding the eventual environmental contamination and operator safety of a formulation may lead to unnecessary drift and hazard. This may well be the case for Nalcotrol, as the data for % volume less than 114 µm indicates that the driftable component of the spray is the same as for water after the formulation has passed through the pump six times. After 1.2 passes through the pump, the driftable component relative to water has been reduced by only 50%.

A second consequence may be misleading information concerning the impact of the spraying process on the formulation. Where a polymer may be designed as a binding agent or as a 'sticker', etc., the high shear stresses in the pumping system may severely degrade the effect that was observed when the formulation was sprayed using a non-shear system such as a pressure can. However, no data is presented in this study to support this conjecture.

As research at LPCAT proceeds in this area, it is becoming obvious that a defined protocol is required for the measurement of droplet spectra of agricultural spray liquids. It is hoped that the work presented here emphasizes this point.

ACKNOWLEDGEMENTS

We would like to acknowledge Hardi and Spraying Systems for supplying us with nozzles; Terra International, DuPont and ICI America for adjuvants and blank formulations. We would also like to thank G.J. Crease for his advice and review.

REFERENCES

[1]   Rogiers L.M. and Bognolo G., "Use of Polymeric Surfactants in Novel Pesticide Formulations Such as Concentrated Emulsions, Suspo-Emulsions and Multiple Emulsions." Pesticide Formulations and Application Systems: 11th Volume, ASTM STP 1112, David G Chasin and Loren E. Bode, Eds., American Society of Testing and Materials, Philadelphia, 1991.

[2]   Lochhead R.Y. and Dodwell R.C., "Polymeric Emulsifiers - A Viable Environmental Alternative for Emulsification of Pesticide Active Ingredients." Pesticide Formulations and Application Systems: 11th Volume, ASTM STP 1112, David G Chasin and Loren E. Bode, Eds., American Society of Testing and Materials, Philadelphia, 1991.

[3]   Bouse L.F., Carlton J.B., and Jank P.C., "Effect of Water Soluble Polymers on Spray Droplet Size." American Society of Agricultural Engineers, Vol. 31(6), 1988, pp. 1633-1648.

[4]   Young B.W., "A Device for the Controlled Production and Placement of Individual Droplets." Pesticide Formulations and Application Systems: Fifth Volume, ASTM STP 915, L.D. Spicer and T.M. Taneko, Eds., American Society for Testing and Materials, Philadelphia, 1986, pp. 13-22.

[5]   May K.R., "The Measurement of Airborne Droplets by the Magnesium Oxide Method." J. Sci. Inst., Vol 27, May 1950, pp. 128-130.

H. Erdal Ozkan, Donald L. Reichard, Kevin D. Ackerman and Jeffrey S. Sweeney[1]

EFFECT OF WEAR ON SPRAY CHARACTERISTICS OF FAN PATTERN NOZZLES MADE FROM DIFFERENT MATERIALS

REFERENCE: Ozkan, H. E., Reichard, D. L., Ackerman, K. D., "Effect of Wear on Spray Characteristics of Fan Pattern Nozzles Made from Different Materials," Pesticide Formulations and Application Systems: 11th Volume, ASTM STP 1112, Loren E. Bode and David G. Chasin, Eds., American Society for Testing and Materials, Philadelphia, 1992.

ABSTRACT: Spray nozzles are important in metering the liquid, atomizing, and controlling the distribution of spray over the spray swath. The effects of nozzle wear on flow rates, spray patterns, and droplet size spectrums delivered by fan pattern nozzles made from different materials were investigated. The relative wear rates of nozzle materials varied greatly with different times of use. At the end of test periods, hardened stainless steel was the most resistant to wear, followed in order of decreasing resistance to wear by stainless steel, plastic, nylon and brass. Nozzles with lower capacities wore at faster rates than nozzles with higher capacities. An automated computerized weighing system was developed to rapidly evaluate the effect of nozzle orifice wear on the spray deposit distribution. The results indicate that there was little difference between the spray deposit distributions of new and worn fan nozzles. The width of the spray pattern remained nearly constant but the worn nozzles delivered greater volumes of liquid in the centers of the patterns. Generally there was little difference in droplet size distributions between new and worn nozzles.

KEYWORDS: nozzle, wear, spray, pattern, patternator

[1]Dr. Ozkan is Associate Professor of Agricultural Engineering at the Ohio State University; Mr. Reichard is Agricultural Engineer, USDA - Agricultural Research Service, Application Technology Research Unit, Wooster, Ohio; Mr. Ackerman and Mr. Sweeney are students in the Agricultural Engineering Department at the Ohio State University.

INTRODUCTION

Applying pesticides properly is essential to achieving satisfactory weed, disease, and insect control.   The directions on a pesticide container label indicate the recommended application rates for best results.   However, proper application can be achieved only if the pesticide application equipment is performing properly, and is calibrated and operated correctly.   Application accuracy is directly affected by how well each component of the sprayer performs.   Spray nozzles, although they are an inexpensive component of a spraying system, have considerable influence on satisfactory pest control. They are important in metering, atomizing and controlling the distribution of spray over the spray swath. Changes in these functions of a nozzle can influence the effectiveness of chemicals applied.

Calibration clinics conducted in Ohio revealed that more than 1/3 of the sprayers surveyed were overapplying chemicals [1].   The major reason for overapplication was worn nozzles.   Agricultural spray nozzles are made from a variety of materials including brass, stainless steel, hardened stainless steel, nylon, plastic and ceramic. Nozzles made from different materials have different wear characteristics.   Other factors which affect nozzle orifice wear include the shape and size of orifice, spraying pressure, usage time, and type of formulation applied.

Some applications require precise overlapping of adjacent nozzle patterns. Litte is known about the influence of nozzle wear on spray distributions. When nozzles wear, they may no longer produce the spray pattern essential for   uniform coverage.   Such distortions in spray patterns could cause streaks of under or over dosed areas. Similarly, nozzle wear may affect droplet size which may be an important factor in achieving satisfactory pest control. Some research has shown that small droplets provide more efficient use of insecticides than larger droplets. However, small droplets are highly susceptible to spray drift.

Nearly all known research on nozzle wear [2,3,4,5,6] has concentrated on the change in flow rate.   Only one known study investigated the change in spray pattern [7]. No known research has investigated the effect of nozzle wear on droplet size. One reason for the lack of research pertaining to effect of nozzle wear on spray pattern is the lack of equipment needed to rapidly and accurately measure and analyze the distribution of spray patterns. Carpenter et al. [8] developed a computerized weighing system to analyze spray distributions.   A weighing system using a load cell was designed to continuously collect and weigh the spray from the nozzle as it was moved sideways across a narrow slot.   The major axis of the nozzle spray pattern was oriented parallel to the direction of travel.   The spray passing through the slot was collected and weighed directly by the load cell.   Although the results obtained with this system were consistent and comparable to patternator results, several problems were identified.   Problems were caused by air movement, spray impact on the collection container, and drive motor vibrations transmitted through the table to the load cell. We decided to develop a new system.

The objective of this study was to investigate the effect of nozzle wear on spray characteristics of fan nozzles constructed with different materials and flow capacities. The spray characteristics investigated were flow rate, distribution pattern, and droplet size.

EQUIPMENT AND PROCEDURE

The first stage in our tests was to measure the change in flow rate as a result of nozzle wear. To study the change in flow rate, a test stand illustrated in Figure 1 was constructed. The test stand consists mainly of a 208 L (55 gal) tank with a pumping system to circulate a mixture of water and an abrasive agent through nozzles. A mechanical, paddle-type agitator at the bottom of the tank was used to ensure a uniform mixture of water and abrasive during nozzle wear tests. With this system, as many as 18 nozzles can be tested simultaneously. In addition to the type of nozzle material, the system was used to study the effect of orifice size on nozzle wear. Fan-pattern nozzles with flow rates of 0.8, 1.5, 2.3, and 3.0 L/min (0.2, 0.4, 0.6, and 0.8 gal/min) were selected for these tests. Nozzle materials included brass, stainless steel, hardened stainless steel, nylon and plastic. The test procedure was the same for each nozzle. A 150 liter tank mixture that contained 60 grams of Hydrite Flat D (Georgia Kaolin Co.) per liter of water was circulated through three nozzles of each size and type. An electromagnetic flowmeter was used to measure the flow rates delivered by the nozzles. Detailed information about the nozzle wear test stand, and procedures used to determine the changes in flow rates of nozzles is given by Reichard et al. [9].

When tests to determine the changes in flow rates were completed, the worn nozzles were used to determine changes in spray distributions due to orifice wear. All spray distribution tests were performed on an automated spray patternator table that we constructed. The patternator consisted of a pumping system, spray table, and an automated weighing and data acquisition system. The spray nozzle was mounted over the patternator table and sprayed onto it while a load cell moved across the bottom of the patternator (Fig.2). A 0.56 kW (3/4 Hp) electric motor is used to drive a roller pump which pumps water from a storage tank to the nozzle. A spray pressure of 276 kPa (40 psi) was maintained throughout the tests. Water discharged from the patternator was returned to the storage tank.

The patternator table consisted of V shaped channels 5 cm (2 in) deep to keep water from splashing out and into adjacent channels, and 4 cm (1.5 in) wide (from peak to peak). The patternator was tilted 5 degrees to direct the water into a collection unit at the end of the table.

The collection unit consists mainly of a beaker and an electronic balance. The beaker and balance are enclosed, except for a narrow slot

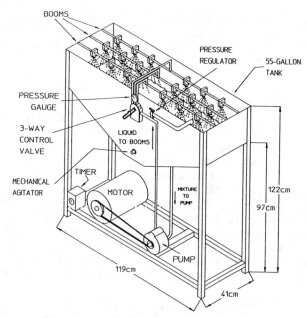

Figure 1. Nozzle wear test stand.

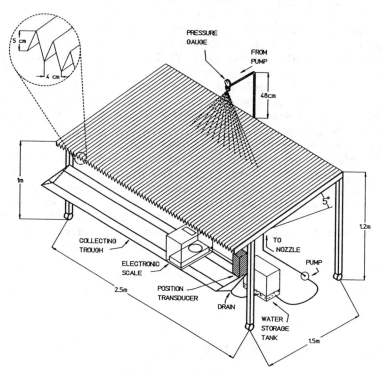

Figure 2. Automated spray nozzle patternator system.

in the top, by a plexiglass housing. This housing prevents spray that is not collected by the beaker from interfering with measurements. Spray can enter the housing through a 4 by 10.5 cm (1.5 by 4 in) slot in the top and deposit in the beaker.

The collection unit is mounted on a chain driven assembly which can move it along the end of the channels. A variable speed 0.37 kW (1/4 hp) motor drives chains which transport the collection unit. For these tests, the collection unit moved at a constant velocity of 9 cm/s. The position of the collection unit is determined by a position sensor attached to the collection unit with a cable. The weight of the water collected in the beaker is measured 7 times /second and sent to an IBM/AT micro computer via an RS-232 cable; collector position data are sent at the same rate, through an analog to digital board to the computer. The data are transferred to the computer for analysis while the collection unit traverses. Graphical representation of the data is accomplished with ASYST software. With this system, it takes approximately one minute from the time the collection unit is activated until the spray distribution data is displayed on the computer screen.

Droplet size distributions were measured with a Malvern 2600C laser droplet and particle sizer. It is capable of measuring droplets ranging in size from 0.5 to 1880 μm with an accuracy of ±4%. The test nozzle was mounted on the horizontal member of an X-Y positioner. Two stepping motors, one for the X positioner and one for the Y positioner, are used to move the nozzle a predetermined distance in either the horizontal (X) or vertical (Y) direction. A controller is used to coordinate the movement of the nozzle by activating the proper stepping motor. The X and Y coordinates of the laser beam position were labeled (0,0) for convenience. The spray nozzle was positioned between the laser transmitter and receiver at a height of 48 cm above the laser beam. This is the height recommended by the nozzle manufacturer for a nozzle with an 80 degree spray angle. Droplet size measurements were taken at 1 cm intervals over -46 cm to 46 cm in the X direction on either side of the laser beam. This X range included the entire spray pattern plus about 5 cm on either side of the pattern. The water pressure at the nozzle was maintained at 276 kPa (40 psi). The time required to complete the test for one data point was about one minute. The total time required to complete all the droplet size measurements for one nozzle required approximately 3 hours. The statistics determined at each test location included the $Dv_{.1}$, $Dv_{.5}$ (Volume Median Diameter), SMD (Sauter Mean Diameter), $Dv_{.9}$, and percent of droplets within each of the 32 size classes.

RESULTS

Effect of nozzle wear on flow rate

Following is a summary of results related to the effect of orifice wear on nozzle flow rate. The detailed results are given by Reichard et al. [9].

Tests to measure the increase in flow rate with wear were stopped when there was about 10% increase in flow rate through stainless steel nozzles. Tests with brass nozzles were terminated before other nozzles because of excessive wear. The brass nozzles wore much more rapidly than the other nozzles (nylon, plastic, stainless steel and hardened stainless steel). Table 1 shows the number of hours the nozzles were subjected to wear tests and the percent flow rate increase at the end of the test period. The values for the percent flow rate increase are the average of three nozzles tested in each group of nozzles made from different materials. Although hardened stainless steel was the most

TABLE 1 -- Duration of wear tests and increase in flow rate of nozzles at the end of the test period

Nozzle Capacity L/min (gpm)	Brass		Nylon	
	Test Period (Hr)	Increase in flow rate (%)	Test Period (Hr)	Increase in flow rate (%)
0.8 (0.2)	26	22.8	40	15.8
1.5 (0.4)	44	19.6	100	17.9
2.3 (0.6)	110	21.0	242	23.8
3.0 (0.8)	268	20.2	352	17.7

Nozzle Capacity L/min (gpm)	Plastic		Stainless Steel	
	Test Period (Hr)	Increase in flow rate (%)	Test Period (Hr)	Increase in flow rate (%)
0.8 (0.2)	40	18.1	40	14.0
1.5 (0.4)	100	12.4	100	12.2
2.3 (0.6)	242	14.0	242	12.6
3.0 (0.8)	352	13.5	352	11.0

Nozzle Capacity L/min (gpm)	Hardened Stainless Steel	
	Test Period (Hr)	Increase in flow rate (%)
0.8 (0.2)	40	1.5
1.5 (0.4)	100	5.5
2.3 (0.6)	242	5.3
3.0 (0.8)	352	10.2

resistant to wear, followed in order of decreasing resistance to wear by stainless steel, plastic, nylon and brass at the end of test period, this order was different at different times of usage. For example, there was less increase in flow rate with plastic tips than with stainless steel tips for some time after starting the tests. However, the increase in flow rate with stainless steel tips was always less than plastic tips at the ends of the tests. For the same material, nozzles with lower flow capacities wore at a much faster rate than nozzles with higher capacities. For all nozzle capacities, the stainless steel tips had average use times 5.6 and 2.1 times longer than brass and nylon tips, respectively, before flow rates increased ten percent. The percent increase in flow rates for all brass, stainless steel and nylon tips varied directly with approximately the square root of time of use.

Effect of nozzle wear on spray pattern

Spray pattern analysis required more repetition of tests than originally planned. Initial tests showed variations in spray patterns of identical nozzles, both new and used, under the same test conditions. To make sure this variability did not come from experimental errors, we repeated the test procedure several times using the same nozzle and compared the results. Although the repeatability of the distribution measurements was good, we repeated the tests three times for each nozzle tested.

Figures 3 and 4 show spray patterns of nozzles with initial capacities of 0.8 and 1.5 L/min (0.2 and 0.4 gal/min) respectively, both when they were new and after they were worn. Each data point on these graphs represents a composite value of nine measurements taken from experiments with three identical nozzles, with each spray pattern measured three times. The results generally indicate that, for the nozzles and amount of wear tested, there was little difference between the shapes of spray deposit patterns of new and worn nozzles toward both ends of the patterns. However, there was greater difference between new and worn nozzles in volumes of liquid collected in the centers of the patterns than at the edges of the patterns. However, it should be noted that Figures 3 and 4 represent different wear rates which are listed in Table 1. We will continue analyzing the variability in spray distributions of other nozzles used in our wear tests.

Effect of Nozzle Wear on Droplet Size

The results of droplet size measurements obtained from new and worn nozzles with nominal flow rates of 0.8 L/min (0.2 gal/min) are shown in Figure 5. Regardless of nozzle material, $Dv.5$ was smallest at the centers of the patterns and increased toward both ends of the patterns. The $Dv.5$ distributions were parabolically-shaped. At 0 cm the $Dv.5$ of the nozzles tested was approximately 150 μm, and it remained relatively constant within ±20 cm of the center of the pattern. Generally there was little difference in $Dv.5$ distributions between new and worn nozzles. Droplet size distributions generated by nozzles with the same flow capacity, but different materials, were approximately the same. We will continue analyzing the effect of orifice wear on droplet size of other nozzles with flow rates of 1.5, 2.3 and 3 L/min (0.4, 0.6 and 0.8 gal/min).

CONCLUSIONS

1. Hardened stainless steel was the most resistant to wear, followed in order of decreasing resistance to wear by stainless steel, plastic, nylon and brass at the ends of test periods.

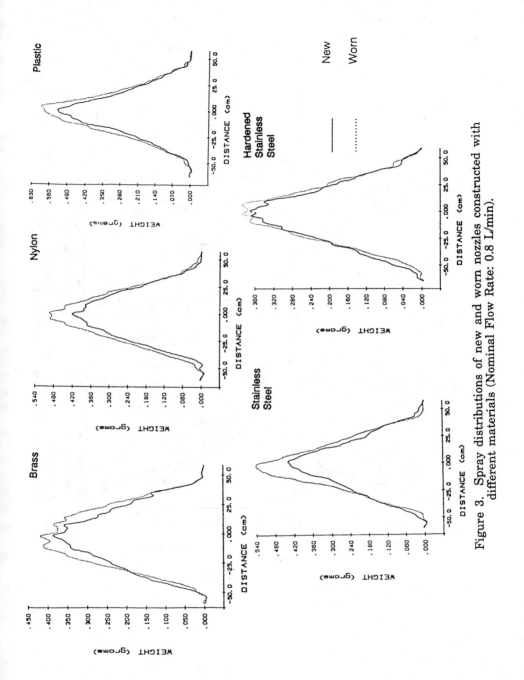

Figure 3. Spray distributions of new and worn nozzles constructed with different materials (Nominal Flow Rate: 0.8 L/min).

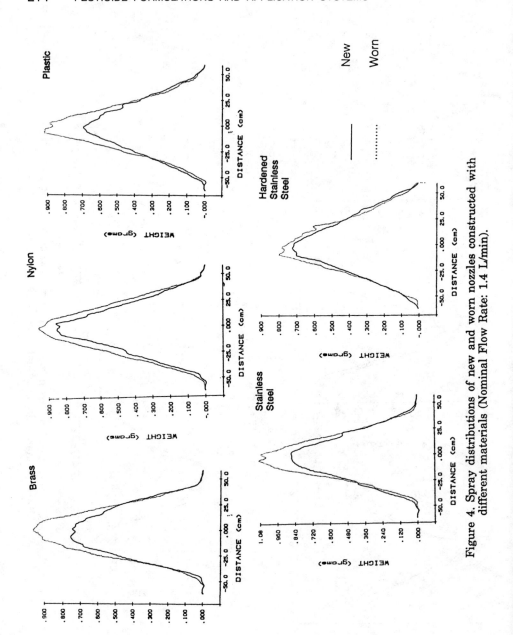

Figure 4. Spray distributions of new and worn nozzles constructed with different materials (Nominal Flow Rate: 1.4 L/min).

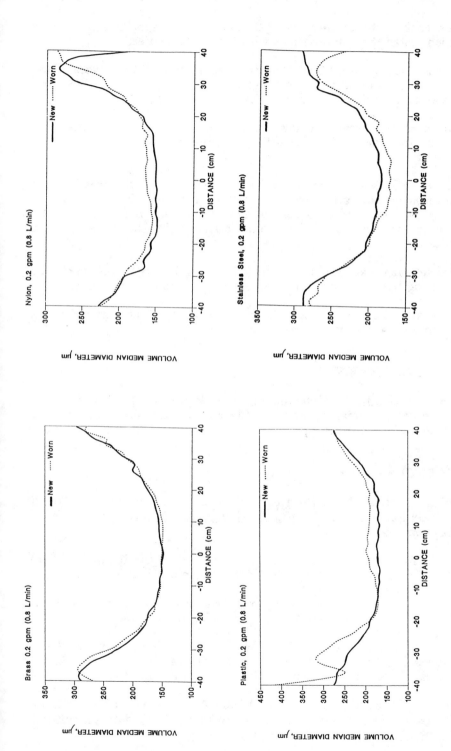

Figure 5. Effect of wear on Dv.5 (Volume Median Diameter) of new and worn nozzles with nominal flow rate of 0.8 L/min.

2. For each material, nozzles with lower flow capacities wore at a much faster rate than nozzles with higher capacities.

3. For all nozzle capacities, the stainless steel tips had average use times 5.6 and 2.1 times longer than brass and nylon tips, respectively, before flow rates increased ten percent.

4. The percent increase in flow rates for all brass, stainless steel and nylon tips varied directly with approximately the square root of time of use.

5. There was little difference between the shapes of spray deposit patterns of new and worn fan pattern nozzles. The width of the spray pattern tended to remain nearly constant but there was an increase in the volume of liquid collected in the center of the pattern.

6. Droplet size distributions generated by nozzles with the same flow capacity, but different materials, was approximately the same.

7. Generally there was little difference in droplet size distributions between new and worn nozzles.

ACKNOWLEDGEMENTS

The authors thank H. Almekinders, H.D. Bauman, D. L. Collins for their technical assistance; and R. Clason for his assistance in the preparation of this manuscript.

REFERENCES

[1] Ozkan, H. E., "Sprayer Performance Evaluation with Microcomputers," Applied Engineering in Agriculture, Vol. 3, No. 1, 1987, pp. 36-41.

[2] Reed, T. F. and Ferrazza, J., "Wear Life of Agricultural Nozzles," ASAE Paper No. AA84-001, ASAE, St. Joseph, MI, 1984.

[3] Friesen, O. H., "Evaluation of Wear Rates of Flat Spray Nozzles," Manitoba Agriculture, Winnipeg, Manitoba, August 1984.

[4] Pearson, S. L. and Fry, C., "Wear Rates of Agricultural Spray Nozzles,"Thirty-sixth Illinois Custom Operators Training School Manual, Cooperative Extension Service, Urbana, IL, 1984.

[5] Menzies, D. R., Fisher, R. W. and Neff, A. E., "Wear of Hollow Cone Nozzles by Suspensions of Wettable Powders," Canadian Agricultural Engineering, Vol. 18, No. 1, 1976, pp. 14-15.

[6] Novak, M.J., and Cavaletto, R.A., "Wear Characteristics of Flat Fan Nozzles. ASAE Paper No. 88-1015, St. Joseph, MI, June 1988.

[7] Doll, J. D., Knake, E. L. and Butler, B. J., "Effect of Wear on Nozzle Tips,"Illinois Research, Spring 1966, pp. 10-11.

[8] Carpenter, T. G., Reichard, D. L., Ozkan, H. E., Holmes, R. G. and Thornton, E. "Computerized Weighing System for Analyses of Nozzle Spray Distribution," Transactions of the ASAE, Vol. 31, No. 21, 1988, pp.375- 379.

[9] Reichard, D. L., Ozkan, H. E. and Fox, R. D., "Nozzle Wear Rates," ASAE Paper No. 901584, ASAE, St. Joseph, MI, December 1990.

Gilbert V. Chambers, Michael C. Bulawa, Chester G.
McWhorter, and James E. Hanks

USE OF SURFACE RELATIONSHIP MODELS TO PREDICT THE
SPREADING OF NONAQUEOUS DROPLETS ON JOHNSONGRASS

REFERENCE: Chambers, G. V., Bulawa, M. C., McWhorter,
C. G., and Hanks, J. E., "Use of Surface Relationship
Models to Predict the Spreading of Nonaqueous Droplets
on Johnsongrass," Pesticide Formulations and Applica-
tion Systems: 11th Volume, ASTM STP 1112, Loren E.
Bode and David G. Chasin, Eds., American Society for
Testing and Materials, Philadelphia, 1992.

ABSTRACT: A renewed interest in improving the application of chemical pesticides has prompted additional
application research on utilizing spray adjuvants or
carrier systems for enhancing ground or aerial applications. This application research is producing
findings on the potential of nonaqueous application
systems as well as research leads for improving pesticide formulations. Nonaqueous application systems
provide greater surface coverage than aqueous or
emulsion systems. While the total spray volume for
nonaqueous carriers was substantially less in field
comparisons, efficacy was better. In comparisons of
nonaqueous carriers, mineral oil was superior to
vegetable oils (soybean, cottonseed) in weed control.
A number of factors can affect the surface coverage of
pesticide formulations. This paper evaluates factors
for relationship to surface coverage on grasses,
including: (1) the surface chemistry of the carrier/
adjuvant and (2) leaf chemistry of grasses. Other
measurements of spreadability on inert surfaces were
compared to the measurement of coverages on johnsongrass leaves.

KEYWORDS: nonaqueous application, mineral oil, vegetable oil, ULV, wetting, leaf surface coverage,
johnsongrass, leaf wax chemistry

Mr. G. V. Chambers and Dr. M. C. Bulawa are research
chemists at Exxon Research and Engineering Co., P. O. Box
4255, Baytown, TX 77522 and Dr. C. G. McWhorter and Mr.
J. E. Hanks are research scientists at USDA-ARS, Southern
Weed Science Laboratory, P. O. Box 350, Stoneville, MS
38776.

Nonaqueous application systems can provide greater surface coverage than aqueous or emulsion systems. While the total spray volume for nonaqueous carriers was substantially **less** in field comparisons, efficacy **can be better**. An increase in efficacy with nonaqueous application has been observed in both insect and weed control investigations [1-4]. Three mechanisms have been attributed to nonaqueous application systems for enhancing the penetration of herbicides: greater surface coverage, partial solubilization (softening and distortion) of waxy platelets [5], or solvency for the herbicide. Wetting of the leaf surface is important for foliar uptake since overall foliar absorption, being a diffusion process, should increase with surface coverage [6]. While coverage can be important in the application of many pesticides, the temporary physical modification of a leaf surface is considered the more important mechanism with nonaqueous application of postemergence herbicides. Early studies revealed that the application of a nonaqueous adjuvant even before the application of a postemergence herbicide was equivalent in weed control to applying them at the same time [2,7,8]. A nonaqueous adjuvant can have a surface tension less than the critical surface tension of the plant surface. While penetration of stomatal pores would be enhanced, the role of the stomata is still uncertain [9].

A leaf surface is a nonuniform solid surface. This nonuniformity occurs in both surface morphology and heterogeneity in a chemical sense. Both the microscopic roughness of the surface and the chemical composition of the epicuticular wax are important in determining wettability of a leaf surface.

The intent of this study was to examine the wetting process and to compare methods for evaluating wetting. A number of factors can affect the wetting and subsequent coverage on the surface of the leaf. Factors evaluated for their relationship to surface coverage on grasses included: (1) the surface chemistry of the carrier/adjuvant and (2) leaf chemistry of grasses. Laboratory measurements of wetting on synthetic surfaces and johnsongrass were compared to measurements of coverage on johnsongrass (*Sorghum halepense*) leaves. While measures of surface roughness were not obtained, a general chemical characterization of the epicuticular wax was made. Epicuticular wax refers to the surface wax including the wax component of trichomes and other specialized structures found at the surface.

MATERIALS AND METHODS

Adjuvant Surface Chemistry

A series of petroleum and vegetable oils were examined by physical and chemical methods which reflect differences in the wetting and movement of adjuvants on a plant surface. Nonionic surfactants were included in the study to provide results for comparing aqueous to nonaqueous. Physical and chemical characterization methods included surface tension at 25±2C (tensiometer method), kinematic viscosity in centistokes (CS) at 37.8C (100F), 25C, and 20C [ASTM D-445: Standard Test Method for Kinematic Viscosity of Transparent and Opaque Liquids (and the Calculation of Dynamic Viscosity)], specific gravity at 15.6C (60F) (ASTM D-4052: Standard Test Method for Density and Relative Density of Liquids by Digital Density Meter). Both midboiling point (MBP) and calculated vapor pressure were obtained from gas chromatographic distillation data (ASTM D2887: Standard Test Method for Boiling Range Distribution of Petroleum Fractions by Gas Chromatography) or literature values. This characterization of test materials is shown as Table 1.

Surface Modeling

Two additional laboratory methods were studied for comparing wetting to coverage on a plant surface. The two laboratory methods were: (1) the movement of droplets down an inclined surface with time as described by McKay [10] and (2) the contact angle (θ) of test materials on synthetic surfaces and johnsongrass. McKay chose a lapse time of five minutes for observing droplet movement because of the high volatility of water. Test materials were evaluated at intervals of time up to one hour since they had a wide range in volatility. Two test runs were performed; each run was comprised of five 10 microliter (μl) droplets. The median value for five droplets was used; McKay used the average of three droplets after discarding the shortest and longest movement. The two median values were averaged.

The synthetic surfaces for comparing contact angle were plastic-coated paper, as used in the droplet movement method, and paraffin wax. Each lot of plastic coated paper was standardized by determining the contact angle of methylene iodide and hexadecane. A paraffin wax (Paravan 138) with a carbon number range of C_{23} to C_{33} was selected. This range of normal paraffins was close to the carbon range of paraffins identified in the surface wax of johnsongrass. A flat specimen of paraffin wax was prepared by hot water casting. Paraffin wax was heated above its melting point of 59C, poured into an 18 x 25 centimeter (cm) pan of water, gently stirred, and allowed to naturally cool overnight. The 0.15 cm thick slab of wax was cut into smaller specimens (1.25 x 5.0 cm). Only the water side of a casting was used, being smoother than the air side.

Contact Angle on Johnson Grass

Measurements of contact angle on field collected johnsongrass were obtained in the same general dorsal area of a leaf blade that had been selected for measuring surface coverage. Leaf blade samples were the fourth to seventh leaf of entire field grown plants. Contact angle measurements were completed within 4 hours for a specific lot of plants.

TABLE 1 -- Characterization of Petroleum and Vegetable Oils and Nonionic Surfactants

MATERIAL	DESCRIPTION	MID-BOILING POINT, F	VAPOR PRESSURE mm Hg AT 100 F	SPECIFIC GRAVITY 15.6 C (AT 60 F)	VISCOSITY, CENTISTOKES AT 37.8 C (100 F)	25 C	20 C	SURFACE TENSION, DYNES/cm, 25 C ± 2
WATER	H_2O, DISTILLATE	212	49.00 (b)	1.000	...	...	...	69.4
METHYLENE IODIDE	CH_2I_2	356 (a)	...	3.325 (20 C) (a)	...	...	...	50.8 (d, 20 C)
HEXADECANE	$CH_3(CH_2)_{14}CH_3$	550 (a)	<1.00 (b)	0.774 (20 C) (a)	3.1	4.1	4.4	27.6 (d, 20 C)
ISOPAR®M	HIGHLY BRANCED ISOPARAFFINS AND ALKYL CYCLOPARAFFINS	430	1.50 (c)	0.781	2.2	3.0	3.3	24.9
ORCHEX®692	PARAFFINIC HYDROCARBONS PREDOMINATELY C_{21}	690	0.0028	0.840	9.7	15.1	18.6	30.3
ORCHEX®796	PARAFFINIC HYDROCARBONS PREDOMINATELY C_{23}	715	0.00011	0.842	11.8	18.6	22.9	29.7
SUN 7N	PARAFFINIC HYDROCARBONS	710		0.861	16.5	29.4	35.6	30.6
CHEVRON AG OIL 100	PARAFFINIC OIL	760		0.857	22.1	40.8	53.8	29.9
CHEVRON VOLCK OIL	98% PARAFFINIC OIL 2% EMULSIFIER	755		0.857	21.2	37.3	48.0	31.1
COTTONSEED OIL	AGRICULTURE GRADE	1100+		0.923	35.9	59.0	74.0	33.5
SOYBEAN OIL	AGRICULTURE GRADE	1100+		0.925	35.1	57.3	77.6	33.6
RAPESEED OIL	AGRICULTURE GRADE (FRANCE)	1100+		0.925	37.7	60.3	74.0	32.9
CANOLA OIL	CONSUMER PRODUCT GRADE	1100+		0.921	39.2	66.2	82.0	33.3
TRITON® X-100	OCTYL PHENOL ETHOXYLATE, 10 MOLES EO	...	...	...	13.5	HLB VALUES		30.7 (0.1 WT%)
STEROX®N.J.	NONYL PHENOL ETHOXYLATE, 6 MOLES EO	...	...	...	13.0			31.0 (0.1 WT%)
T MULZ®AO2	BLEND: DODECYLPHENOL ETHOXYLATE, 8 MOLES EO	...	...	...	11.3			...
EXXON AROMATIC 150	C_{9-10} ALKYL BENZENES	415	2.50 (c)	0.895	1.1	1.3	1.3	30.7
EXXON AROMATIC 200	C_{11-12} ALKYL NAPHTHALENES	465	1.30 (c)	1.003	2.0	2.6	3.0	37.4

(a) SOURCE: LANGE'S HANDBOOK OF CHEMISTRY
(b) SOURCE: PERRY, J. H., CHEMICAL ENGINEERS HANDBOOK
(c) ASTM D 2879
(d) SOUHENG, W. U., POLYMER INTERFACE AND ADHESION, 1982, p. 176.

Each lot of plant material was standardized by determining the contact angle of methylene iodide and hexadecane or Orchex® 796. While rhizome johnsongrass with seed heads was used in both methods, our johnsongrass materials were not from the same field nor did they have an identical environmental history.

Contact angle was obtained on both the left- and right-hand side of three droplets; the average value was reported. Contact angle was measured with a NRL C.A. goniometer. Drops were formed by discharging a liquid from a 1 milliliter (ml) disposable syringe through a 23 GHW needle. The volume of a droplet was 6 to 8 μl.

Both linear regression analysis by parametric statistics and correlations by nonparametric statistics were performed. Contact angle values of less than 3 degrees were assigned a value of 2 degrees for statistical analysis.

Spread Coefficients on Johnsongrass

The determination of a spread coefficient on a johnson-grass leaf for a 2-μl droplet was described earlier [1]. The spread coefficients in the current studies are based on a 1-μl droplet and the average from two experiments. Each experiment had 12 determinations of the radius of the spread on leaf blades of a known age. Measurements on spread were taken after the spread of liquids had ceased. This time period, between application and recording spread, varied from 5 to 10 minutes for water mixtures, to 15 to 45 minutes for lower molecular weight oils, to 3 to 4 hours for Orchex 796 and higher molecular weight materials. The three lowest and the three highest values were discarded. A spread coefficient was calculated from this averaged radius of the spread as defined below.

$$\text{Spread coefficient} = \frac{(r_2)^2}{(r_1)^2} \text{ , where}$$

r_1 = the radius of a 1 μl droplet, r_2 the radius of area of spread on the surface of a johnsongrass leaf.

Both test materials and air temperature were 22±1C.

Leaf Surface Chemistry

Epicuticular wax was collected from leaf blades of field and greenhouse johnsongrass plants. Leaf blades were cut from plants at their base, measured to determine surface area, and washed in a metal pan with high pressure liquid chromatograph (HPLC) grade chloroform for 30 seconds. Chloroform solutions were filtered through chloroform washed analytical grade filter paper, transferred to glass flasks, and evaporated at 22C. A stream of filtered air was directed over the surface of the chloroform to increase the rate of evaporation.

After about 90% evaporation had occurred, the solution was transferred to a smaller glass container and evaporation continued to dryness. Wax samples were stored at about -15C prior to analysis. Filaments of the leaf blade were collected by brushing the leaf with a brush which was wet with ethanol. Filaments were stored in ethanol.

The characterization of epicuticular wax has been previously carried out by other investigators. This study is an adaptation of procedure developed by Tulloch [11] who converted the free alcohols and acids into their trimethyl-silyl ethers and esters. The waxes were received as dried chloroform extracts and were analyzed as received. The filaments were filtered from the ethanol and analyzed.

The waxes were trimethlylsilylated as follows: into a 1.0 ml reaction vial, equipped with a teflon backed silicon septum, a sample of wax [20.0 milligrams (mg)] was dissolved in 0.1 ml heptane and 50 μl hexamethyldisilazine, 10μl pyridine, and 0.50mg of normal tetradecanol were added. The vial was sealed and the mixture was then kept at 80C overnight.

The wax components were identified by gas chromatography-mass spectrometry (GC/MS). A Hewlett Packard Model 5970 Mass Selective Detector with a Hewlett Packard Model 5890 gas chromatograph was used. Transfer lines and injection port were kept at 300C. A 15 meter (m) x 0.18 millimeter (mm) inner diameter (i.d.) fused silicon capillary column containing a film of 0.2 micron (μ) methyl silicon was used and inserted directly into the ion source which was held at a temperature of 200C and a pressure approximately $1 \times 10^{(-5)}$ torr. Hydrogen was the carrier gas. The gas chromatograph was temperature programed from 35C at a rate of 5C/minute (min) to 300C and held for 30 min. One microliter of the wax mixture was injected into the injection port at a set for a split ratio of approximately 50:1. The GC/MS was set to scan a mass range of 50-700 atomic mass units (amu).

The percent recovery of the wax components was determined by gas chromatography-flame ionization detection (GC/FID). A Hewlett Packard Model 5890 Series II gas chromatograph with a flame ionization detector, a temperature programmable on column injection port, and a 15m x 0.53mm i.d. aluminum clad fused silica capillary column containing a 0.15μ film of methyl silicon was used. The gas chromatograph held at initially at 100C and then programed to 200C at a rate of 5C/min and held for 15 min. The temperature was increased to 400C at a rate of 10C/min where it was held for an additional 30 minutes. Helium was used as a carrier gas at a velocity of 20cm/second (sec). The detector was operated at 400C. The on-column injection port was temperature programed from 100C at a rate of 30C/min to 400C. A 1μl sample of the wax mixture was injected.

The amount of each individual component was calculated from the response relative to the internal standard n-tetradecanol by integrating the peak areas created from the total ion current from the GC/MS experiments.

RESULTS AND DISCUSSION

Seventeen test materials were compared on the basis of droplet movement or contact angle data (Table 2). The three surfaces for contact angle measurements provided a consistent pattern for all of the nonaqueous test materials. The wax surface had the highest value and the plastic-coated paper had the lowest value. The plastic-coated paper method provided the least distinction between test materials. Nonaqueous test materials with the greater droplet movement did have the expected lower contact angle with two exceptions. The higher density of methylene iodide and Aromatic 200 did apparently affect droplet movement; and the distance traveled at five minutes was greater than reflected by contact angle. In contrast to paper, a wax surface provided a wide range of θ values. Also, sample preparation for the wax was easier than for johnsongrass.

Two known variables in the surfaces were (1) lot-to-lot variability in the 1000 ft rolls of plastic coated paper and (2) johnsongrass field samples. Johnsongrass samples "A" and "B" were from the same field, with the sample labeled "B" being collected 20 days later and after heavy rainfall. Johnsongrass "C" samples were collected from a different field. Test materials can still be grouped or ranked even with the known field variability for johnsongrass.

For both the parametric and nonparametric statistical comparisons, those values shown in Table 2 with an asterisk were combined into the population of values shown in the first column for each of the wetting methods. An example is, the values 10.4 and 6.9 shown in lot 2 were placed into the lot 1 population for statistical analysis purposes.

The comparisons of wetting methods to coverage on johnsongrass are based on nine test materials (Table 3). Seven of the nine test materials were nonaqueous. Both the johnsongrass contact angle values and wax contact values were associated with johnsongrass coverage. Spearman rank correlation (r_s) is significant at 0.05 level. Contact angle is considered a fairly reliable indicator of wetting [12]. Additional results are discussed below by each type of wetting method studied.

Droplet Movement

Movement of a droplet is related to wetting for aqueous droplets [10]. The distance traveled down an incline is also influenced by both volatility and density of the droplet.

TABLE 2 — Comparison of Wetting Methods

TEST MATERIAL	DROPLET MOVEMENT PLASTIC-COATED PAPER, cm IN 5 MIN		CONTACT ANGLE, θ									
			PLASTIC-COATED PAPER		PARAFFIN WAX				JOHNSONGRASS		
	LOT 1	LOT 2	LOT 1	LOT 2	DAY 1	DAY 2	DAY 3	DAY 4	A	B	C
WATER	0	0	89	-	76	85	82	85	81	85	88
TRITON X-100, 0.1 WT% (0.5)	10.3	8.3	40	-	-	49	49 (40) (a)	-	28	35	-
STEROX NJ, 0.1 WT% (0.5)	-	10.4 (a)	25 (47)	23 (54)	-	-	42 (54) (a)	-	-	31 (41) (a)	-
CH2I2	50+	-	38	43	51	46	50	57	43	43	42
HEXADECANE	14.7	13.8	<3	<3	4	-	<3	<3	-	<3 (a)	<3
ORCHEX 796	9.5	9.1	<3	-	22	-	-	21	8	<3	<3
ORCHEX 692	10.4	-	<3	-	27	-	-	-	10	-	<3
ISOPAR M	10.0	-	<3	-	16	-	-	-	<3	-	<3
SUN 7N	8.0	-	6	-	33	-	-	-	13	-	7
VOLCK OIL	7.3	6.8	4	-	33	-	-	-	13	-	-
CHEVRON AG OIL 100	-	6.9 (a)	6	9	-	-	26	31 (a)	-	-	13 (a)
COTTON SEED OIL	5.8	-	5	-	37	-	-	-	27	-	26
SOYBEAN OIL	5.9	-	8	-	44	-	-	-	27	-	27
RAPESEED OIL	6.0	-	7	-	40	-	-	-	26	-	-
CANOLA OIL	5.5	-	<3	-	39	-	-	-	-	23 (a)	-
EXXON AROMATICS 150	13.0	-	<3	-	-	11	-	16 (a)	-	<3 (a)	-
EXXON AROMATICS 200	21.2	-	<3	-	-	22	-	29 (a)	-	9 (a)	-

(a) - VALUES COMBINED INTO THE POPULATION AS SHOWN IN THE FIRST COLUMN FOR STATISTICAL ANALYSIS PURPOSES.

TABLE 3 — Spread Coefficients of 1-μl Droplets Following Application to the Upper Surface of Johnsongrass Leaves

TEST MATERIAL	r_1^2/r_2^2 AT VARIOUS AGE OF LEAVES		
	EXP-1, 30 ± 5 DAYS	EXP-2, 28 ± 7 DAYS	EXP-2, 56 ± 7 DAYS
DISTILLED WATER	1	0.8	0.8
STEROX NJ, 0.1 V/V%	3	...	...
STEROX NJ, 0.5 V/V%	3	...	...
ORCHEX 796	279	5	73
ORCHEX 692	261	10	124
ISOPAR M	148	89	115
SUN 7N	136	4	4
CHEVRON AG OIL 100	183	...	...
COTTONSEED OIL	17	3	3
SOYBEAN OIL	3	3	3

TEMPERATURE OF STUDIES WERE 72° ± 2 °F
V/V%: VOL. TO VOL. PERCENT OF SURFACTANT TO WATER

Comparisons of surface wetting with time, as reflected by the movement of droplets, reveal a greater movement by one hour with nonaqueous treatments because of the volatile nature of water (Table 4). Further, most of the distance traveled by an aqueous drop in five minutes occurs during the ten seconds at the vertical position rather than during the remaining five minutes at the 25 degree incline. A vertical position is used to overcome inertia. Droplets with a higher density, such as methylene iodide and Exxon Aromatic 200, move further than other less dense droplets with equal or less contact angle.

Droplet movement at five minutes and contact angle on the same paper were not associated when all of the test materials were included. Droplet movement and contact angle were associated when based only on nonaqueous treatment (r_s: significant). Most of the remaining deviation was due to the canola values where droplet movement was less than expected with a contact angle of less than 3 degrees. The ranking of treatments by droplet movement at five minutes was not correlated with surface coverage when aqueous treatments again were included, but the association was significant for the nonaqueous treatments. Methylene iodide values were not included because of its high density and both aromatic values were excluded.

Surface wetting is affected by the type of plastic-coated paper. A local purchased plastic-coated paper (butcher paper) was slower than a sample of paper as provided by McKay. A difference between the two samples of plastic-coated paper was even evident in visual appearance; McKay's paper had a shinier and slicker appearance. Triton X-100 at 0.1 wt% and hexadecane were compared on the two types of paper (Table 5). Both test materials were slower on the locally purchased paper. Likewise, lot 2 was slower, less distance than lot 1; contact angle was higher, which agrees with the droplet movement data.

Contact Angle on Three Surfaces

Paravan 138 was selected for comparing contact angle to the surface coverage. This material has a carbon number range similar to the surface waxes identified on johnsongrass. In contrast to recovered leaf surface wax with its predominance of odd-numbered normal paraffins, the wax for the contact angle study has an equal distribution of odd and even numbered normal paraffins. Gas Chromatographic Distillation (GCD) results by carbon number were as follows: C_{23}, 4.0 wt%; C_{24}, 7.8; C_{25}, 9.6; C_{26}, 10.3; C_{27}, 9.3; C_{28}, 8.4; C_{29}, 8.0; C_{30}, 7.2; C_{31}, 6.3; C_{32}, 5.0; C_{33}, 3.6; other normal paraffins are a total of 10 wt% and other paraffinics are 10 wt%.

The strongest linear relationship between any two methods of wetting or an individual method of wetting and surface coverage was the θjohnsongrass and θwax relationship (Table 6).

TABLE 4 — Comparisons of Surface Wetting on Plastic-Covered Paper with Time

TEST MATERIAL	TIME OF TEST, MIN	DROPLET DISTANCE, CM.				
	0.16	5	15	30	60	
0.1 WT% TRITON X100, DISTILLED WATER	9.3	10.3	9.8	10.9	a	
SOYBEAN OIL	2.6	5.9	9.1	12.1	15.8	
COTTONSEED OIL	2.4	5.8	9.0	11.3	14.6	
ORCHEX 796	4.7	9.5	14.0	16.6	21.6	
HEXADECANE	7.5	14.7	17.8	17.4	19.8	
EXXON AROMATIC 150	10.5	13.1	b	...	...	
EXXON AROMATIC 200	8.9	21.2	25.7	28.5	...	

a. EVAPORATED
b. EVAPORATED, WIDE TRACK.

TABLE 5 — Comparisons of Surface Wetting
by Type of Plastic–Coated Paper

A. McKAY PAPER VS. LOCAL PAPER

| | DROPLET DISTANCE AT 5 MIN., cm | | | |
| | McKAY PAPER | | LOCAL PAPER | |
TEST MATERIAL	Avg.	Median	Avg.	Median
TRITON X–100, 0.1 WT%	11.8	11.7	10.7	10.3
HEXADECANE	15.3	15.6	14.5	14.7

B. LOT 1 LOCAL VS. LOT 2 LOCAL

| | DROPLET MOVEMENT OF HEXADECANE, cm | |
TIME LAPSE, MIN.	LOT 1	LOT 2
5	14.0	12.6
15	16.2	13.5

| | CONTACT ANGLE, θ, WITH VARIOUS LIQUIDS, TWO EXPERIMENTS | | | |
| | EXP–1 | | EXP–2 | |
TEST MATERIAL	LOT–1	LOT–2	LOT–1	LOT–2
CH_2I_2	37.7	43.3	32.0	38.8
HEXADECANE	...	...	<3	<3
WATER	81.7	...	78.3	81.5
CHEVRON AG 100	...	...	5.5	8.7

TABLE 6 — Summary of Linear Regression Relationships

Y	X	N	R^2	COMMENTS
MOVEMENT	θ PAPER	16	0.228	CH_2I_2 EXCLUDED FROM ALL COMPARISONS
MOVEMENT	θ PAPER	13	0.343	EXCLUDED AQUEOUS
MOVEMENT	θ PAPER	11	0.369	EXCLUDED AQUEOUS
MOVEMENT	S	13	0.222	$S = \gamma_{LV}(COS\ \theta - 1)$
θ JG	θ WAX	16	0.876	
θ JG	θ WAX	13	0.819	EXCLUDED AQUEOUS
θ JG	θ WAX	13	1.000	Y = 0
θ JG	θ WAX	11	0.820	EXCLUDED AQUEOUS AND AROMATICS
COVERAGE-30	MOVEMENT-5	9	0.307	
COVERAGE-30	MOVEMENT-5	7	0.668	EXCLUDED AQUEOUS
COVERAGE-30	MOVEMENT-15	9	0.427	15 MIN. LAPSE TIME
COVERAGE-30	MOVEMENT-30	9	0.349	30 MIN. LAPSE TIME
COVERAGE-30	θ WAX	9	0.477	
COVERAGE-30	θ WAX	7	0.514	EXCLUDED AQUEOUS
COVERAGE-30	θ WAX	8	0.524	EXCLUDED ISOPAR M
COVERAGE-30	θ WAX	6	0.920	EXCLUDED AQUEOUS AND ISOPAR M
COVERAGE-30	COS θ WAX	9	0.395	
ℓN COVERAGE-30	θ WAX	9	0.727	ℓN = NATURAL LOGARITHM
COVERAGE-30	θ JG	9	0.436	
ℓN COVERAGE-30	θ JG	9	0.701	
COVERAGE-30	LIQUID VIS, 25C	9	0.030	
COVERAGE-30	γ_{LV}	9	0.206	
COVERAGE-56	θ WAX	7	0.458	ONLY 7 COVERAGE VALUES AVAILABLE
ℓN COVERAGE-56	θ WAX	7	0.709	
COVERAGE-56	θ JG	7	0.355	
COS θ WAX	γ_{LV}	16	0.801	γ_{LV} = SURFACE TENSION OF LIQUID
COS θ WAX	γ_{LV}	13	0.560	EXCLUDED AQUEOUS
COS θ JG	γ_{LV}	16	0.941	
COVERAGE-30	θ WAX AND VIS, 25 C	9	0.532	
COVERAGE-30	θ WAX AND VIS, 25 C	6	0.961	EXCLUDED AQUEOUS AND ISOPAR M
COVERAGE-30	θ WAX AND MBP	9	0.508	
COVERAGE-30	MOVEMENT-15 AND VIS, 25 C	8	0.802	STEROX NJ EXCLUDED

Zisman plots were prepared for the wax surface and johnson-grass surface to define the critical surface tension (γc) value for these surfaces [13]. Critical surface tension was 21 and 28 at 23±2°C for wax and johnsongrass, respectively (Figure 1). Critical surface tension equals the surface tension of the liquid which just exhibits a zero contact angle on that solid. Shafron and Zisman noted that surfaces having similar chemical compositions have similar surface tensions. Both the effects of chemical constitution (Parachlor) and molecular packing affect critical surface tension. The size of the difference between γc wax and γc johnsongrass appears reasonable in view of consituents other than -CH3 and -CH2-at the surface of johnsongrass leaves. A paraffin wax (undefined chemically) value of 23 and a hexa-triacontane (C_{36}) value of 21 were reported by Fox and Zisman [13,14]. Polyethylene, having a critical surface tension of 31 at 20C, may be a suitable surface for evaluating non-aqueous or aqueous materials.

While θwax values were associated with coverage (r_s, significant at .01 level), removal of the more volatile adjuvants from the linear relationships did increase R^2. The variable of volatility was included in evaluations by multiple linear regression. Neither inclusion of the midboiling or viscosity at 25C improved R^2. Most of the deviation upon comparing θwax to coverage was due to Isopar M where the lowest θwax value did not provide the greater coverage.

An exponential curve (natural logarithmic) for representing surface coverage improved the curve fitting with both θwax and θjohnsongrass values. The ln(coverage 30) = m(θwax + b is illustrated in Figure 2. θwax provided a higher relationship than θjohnsongrass for ranking coverage.

Surface tension and polarity of solid polymers has been calculated by the harmonic-mean equation which uses the contact angle of water and methylene iodide [13]. Similar calculations were made for the johnsongrass samples. Results are compared below:

	Johnsongrass Samples		
	A	B	C
Solid surface tension (γ), dynes/cm	40.5	39.2	39.2
Polarity, %	24.4	19.8	16.5

The meaning of the decline in the polarity component of the solid surface tension of johnsongrass (γjg) is not known, but the trend resulted from the trend in the water contact angle values. Again, these polarity values are much greater than reported for a normal paraffin surface.

Both θwax and θjohnsongrass were determined at various Orchex 796: water ratios, thereby simulating nonaqueous and low water ratios as well as conventional water ratios for ground application. θwax and θjohnsongrass measurements are compared below.

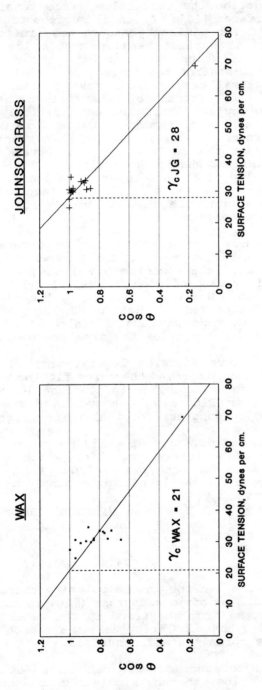

FIGURE 1 -- Zisman Plots for Wax and Johnsongrass Leaves Using Various Liquids to Determine Critical Surface Tension, γ_c

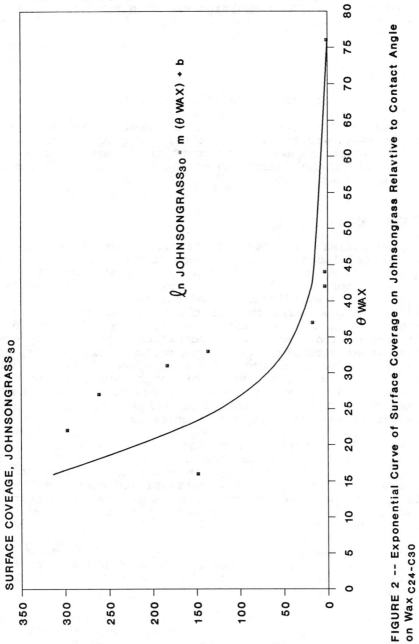

FIGURE 2 -- Exponential Curve of Surface Coverage on Johnsongrass Relavtive to Contact Angle on Wax C24-C30

Change in Contact Angle with Oil:Water Ratio

Oil:Water Vol:Vol	% Orchex 796* in emulsion	Contact Angle θ Wax	θ JG
1:0	100	18.0	<3
1:3	25	37.2	42.8
1:19	5	54.8	48.7
1:79	1.25	73.5	49.0
0:1	0	84.9	-

* 95%; T Mulz AO2 emulsifier at 5 wt%

These results reflect prior field observations and leaf coverage measurements [15]; that is, nonaqueous application (1:0) provides a higher spread coefficient than emulsions or water. Contact angle declined as the level of oil increased in the oil:water ratio.

Leaf Chemistry

Epicuticular wax refers to the surface wax including the various specialized structures found at the surface. The most frequently encountered constituents of plant epicuticular waxes are normal alkanes, primary alcohols, secondary alcohols, ketones, diketones, and esters [16,17]. With our GC/MS analysis we were able to characterize the chloroform extracts of johnsongrass leaf blades as a mixture of predominantly odd carbon number C_{23}-C_{35} normal paraffins, C_{26}-C_{32} even carbon number alcohols, C_{16}-C_{34} even carbon number acids, and four unidentified compounds. The even alcohol chain lengths of C_{26}-C_{32} and the surface waxes of johnsongrass are similar to the waxes of panacoid grasses [18]. Alcohol chain lengths for the tribe Triticeae are almost always C_{26} or C_{28}, very occasionally shorter, but never longer [19].

Figure 3 illustrates a typical GC/MS analysis. Other components can be seen in the chromatogram of Figure 3; however, a positive identification of these constituents can not be made at this time. The unidentified compounds typically contain less than 40 carbons and comprise approximately 15% of the total area in the chromatogram. Ketones and diketones were not detected [11].

Table 7 summarizes the relative distribution of the normal paraffin, linear alcohol and linear fatty acid content of various chloroform extracts. Table 8 describes these extracts.

Because of the capillary splitter injection port on the GC/MS and the presumed nonvolatility of leaf wax components, percent recovery experiments were performed using on-column high temperature gas chromatography with flame ionization detection. This technique is suitable for the analysis of hydrocarbons containing 40 or greater carbons.

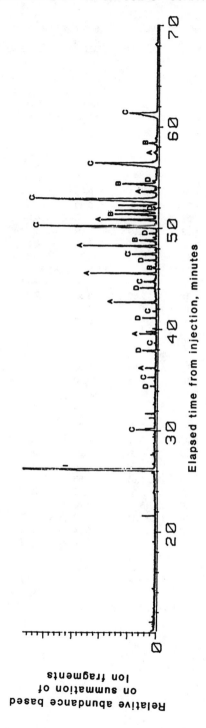

FIGURE 3 -- Total Ion Chromatogram of Sample #2.

Peak Identification:

A = Odd number linear paraffins (C23-C37); B = Even number linear alcohols (C26-C34);

C = Even number linear acids (C16-C34); D = Even number linear paraffins (C22-C36);

I = Internal Standard: tetradecanol.

TABLE 7a — GC/MS Analysis of Johnsongrass
Leaf Blades Chloroform Extracts (1–5)

Compound	RT(min)	1 Dried Johnson %	2 Fresh Johnson %	3 Grass w/o Heads %	4 Grass w/ Seeds %	5 Grass Drought %
Sample Weight (mg)		17.2	13.5	8.5	15.6	31.2
NC23	36.26	0.79	0.74	0.00	0.89	0.54
NC25	39.57	1.34	1.16	0.00	0.34	0.26
NC26	41.14	1.34	1.03	0.00	3.00	1.30
NC27	42.7	5.07	2.91	1.63	4.94	4.85
NC28	44.14	4.09	1.19	0.00	1.98	4.14
NC29	45.61	9.86	5.54	6.85	10.35	7.67
NC30	46.88	1.59	0.92	0.00	1.80	1.52
NC31	48.32	10.44	6.25	9.91	11.72	7.08
NC32	49.54	0.99	0.69	0.00	1.12	1.70
NC33	50.90	7.31	5.03	3.62	7.35	5.24
NC35	53.62	2.52	1.73	1.59	4.42	2.71
NC37	57.27	0	0	0	0	0.83
C24OH	43.29	0	0	0	0	0
C26OH	46.11	0	0	0	0	0.54
C28OH	48.85	1.88	1.94	3.23	2.59	2.06
C30OH	51.39	3.10	4.33	10.60	5.56	3.78
C32OH	54.33	2.76	4.28	48.21	7.79	4.05
C34OH	58.34	0	0	0	0.42	0.91
C16 ACID	31.76	1.53	0.84	0.75	2.81	0.33
C18 ACID	35.36	0.75	0.47	0.00	1.27	0.00
C20 ACID	38.55	0.00	0.00	0.00	0.00	0.00
C22 ACID	41.81	0.50	0.00	0.00	0.96	0.74
C24 ACID	44.74	2.78	1.01	0.65	1.44	4.85
C26 ACID	47.49	2.97	2.68	0.98	2.30	3.73
C28 ACID	50.21	13.36	15.65	3.23	8.65	12.09
C30 ACID	52.83	14.44	22.13	5.72	10.50	14.17
C32 ACID	56.31	7.12	12.57	3.04	6.04	9.87
C34 ACID	61.8	3.47	6.89	0.00	1.74	5.05
Total Paraffins		45.34	27.20	23.60	47.92	37.85
Total Alcohols		7.73	10.56	62.04	16.37	11.33
Total Acids		46.93	62.24	14.37	35.71	50.82

TABLE 7b -- GC/MS Analysis of Johnsongrass
Leaf Blades Chloroform Extracts (6-9) and Filaments (10)

Compound	RT(min)	6 500 Leaves No Seed %	7 500 Leaves w/ Seed Head %	8 Johnson Grass w/o Seed Head %	9 Johnson Grass Harvest 10/16 %	10 Grass Filaments %
Sample Weight (mg)		29.4	16.6	33.3	35.9	7.3
NC23	36.26	0.97	0.85	0.42	0.46	0.08
NC25	39.57	1.99	2.06	1.37	1.83	0.37
NC26	41.14	2.00	2.42	1.25	1.41	0.31
NC27	42.7	4.71	5.95	4.20	3.20	0.20
NC28	44.14	1.68	2.41	1.61	1.90	0.21
NC29	45.61	9.20	10.71	8.76	7.61	0.25
NC30	46.88	1.15	1.38	1.41	2.16	0.08
NC31	48.32	6.82	7.61	6.78	7.59	0.15
NC32	49.54	0.00	0.80	0.48	0.74	0.00
NC33	50.90	2.13	3.04	1.19	2.38	0.23
NC35	53.62	1.37	1.55	0.25	1.59	0.00
NC37	57.27	0	0	0	0.54	0.00
C24OH	43.29	0	0	0.23	0.67	0.00
C26OH	46.11	0	0	1.17	2.54	0.00
C28OH	48.85	1.36	1.23	4.23	9.13	0.16
C30OH	51.39	6.87	5.13	9.10	10.13	0.42
C32OH	54.33	34.43	19.27	36.12	26.20	0.40
C34OH	58.34	0	0	0.15	0.34	0.00
C16 ACID	31.76	2.68	3.89	0.55	0.32	0.31
C18 ACID	35.36	1.87	2.74	0.44	0.26	0.14
C20 ACID	38.55	0.81	1.04	0.26	0.67	0.00
C22 ACID	41.81	1.22	1.52	1.24	1.56	0.00
C24 ACID	44.74	1.81	2.44	2.70	3.46	0.00
C26 ACID	47.49	0.00	0.83	1.78	2.13	0.00
C28 ACID	50.21	2.39	4.61	5.93	4.70	14.90
C30 ACID	52.83	6.17	9.15	4.28	3.20	67.21
C32 ACID	56.31	5.98	6.16	3.72	2.88	12.94
C34 ACID	61.8	2.40	3.23	0.37	0.40	1.63
Total Paraffins		32.02	38.77	27.72	31.41	1.89
Total Alcohols		42.65	25.63	51.00	49.02	0.98
Total Acids		25.33	35.60	21.28	19.58	97.13

TABLE 8 — Description of Johnsongrass Chloroform Extracts (1–9) and Filaments (10)

TABLE 7 NO.	DESCRIPTION OF JOHNSONGRASS	APPROX. AGE- WKS.	NO. OF LEAVES ON PLANT	HEIGHT OF PLANTS, FT.	GROWN IN FIELD (F) GREENHOUSE (G)	NUMBER OF LEAVES	LEAF AREA, cm²	FRESH WT. GMS	EXTRACT WT. GMS
1	AIR DRIED FOR 7 DAYS, (a) WITH SEED HEADS, SEPT. 1988 (92088-2)	18	7-9	6.5	F	200	32,200	340 149 g DRY WT	0.1284
2	WITH SEED HEADS, SEPT. 1988 (92088-1)	18	7-9	6.5	F	200	32,700	36.3	0.2052
3	YOUNG, WITHOUT SEED HEADS (82489A) AUG 15, 1989	7	5-7	3.5 ± 0.5	F	500	108,600	1131.8	...
4	WITH MATURE SEED HEADS (82489B) AUG 15, 1989	13	8-9	6 ± 1	F	500	88,600	1025.0	...
5	SEVERE DROUGHT STRESS, WITH SEED HEAD, 5-6 FT TALL (91688-2)	14	7-9	4.5 ± 1	F	800	...	...	2.0458
6	WITHOUT SEED HEADS (NO. 16)	6	6-7	3.5	G	500	...	...	0.4298
7	WITH SEED HEADS (NO. 17)	12	8-9	6.5	G	500	...	...	0.5851
8	WITHOUT SEED HEADS (NO. 18)	7	7-8	3.5	G	...	...	...	0.5941
9	PLANTED 9-7-89 IN GREENHOUSE HARVESTED 10-16-89 (NO. 19)	6	5-6	3.5	G	...	...	...	1.3468
10	FILAMENTS, FROM PLANTS WITH SEED HEADS	10	8	6.0	G	...	...	...	...

(a) ALL OTHER SAMPLES ARE EXTRACTED FROM FRESHLY HARVESTED PLANTS

As configured, the GC/MS was limited to compounds with molecular weights less than tetracontane. The on-column capillary analysis is illustrated in Figure 4 and reveals the presence of a higher boiling fraction of wax. In terms of carbon number, this fraction would contain compounds of greater than 40 carbons. Previous investigators have reported the presence of the higher molecular weight esters stating that $C_{40}-C_{56}$ esters are a significant component in surface wax of grasses [whole plants, 18,20-23]. Kurtz showed an increase in the ester level with age on johnsongrass leaves [24]. This high boiling fraction, referred to in Figure 4 as the C40+ fraction constitutes approximately 60% of the total peak area in the chromatogram and is most likely the ester fraction. Additional experiments are now underway to confirm the identity of these compounds. The identified fraction containing the lower molecular weight acids, paraffins and alcohols represented approximately 40% of the total peak area in all the samples analyzed. Percent recoveries were calculated from the known weight of the trimethylsilyl ether of tetradecanol and the integrated peak areas. Recoveries ranged from 100 to 110% as determined by this approach.

Samples were selected for analysis to reflect the age of johnsongrass plants. Many workers have reported a change in the total amount of wax or the chemical constitution with age or climate. The trend in the changes in the surface chemistry with age was similar for two sets of johnsongrass leaf samples. Both the normal paraffin and acid component levels increased with age; alcohols declined (Table 9). Drought stressed plants had the highest percent of acid compounds (Samples 2 and 5 in Table 9). Young grass (Sample 3) and mature grass (Sample 4) are compared to drought stressed (Sample 5) in Figure 5. Samples 3 and 4 were collected in 1989 from the same field at the same age as Samples 1 and 2 in 1988. Drought stress was greater in 1988.

The relative abundance of C_{28}, C_{30}, or C_{32} alcohols may reflect the effects of growing conditions. C_{32} alcohol level declined over the drier '88 period. C_{26}, C_{28}, C_{30} alcohols were normalized to define percent primary alcohol by carbon number (Table 9). Tulloch, et al, 1980 found alcohol chain length on whole plants was rarely affected by growing conditions, but occasionally the varieties of a species did differ in chain lengths. Samples 3 and 4 reflect the decline in C_{32} alcohol and both samples were from the same field and are considered the same variety of johnsongrass.

The increase in acid content of the extract with time and with age is likely related to the amount of filaments present. Filaments are produced later in the life of a leaf [25]. Filaments were isolated from johnsongrass, analyzed, and found to consist of 97% linear fatty acids, mostly n-tricontanoic acid (Figure 6). If the filaments are a major constituent for the observed increase in acid content of the chloroform extracts with age and drought, the paraffin content of the remainder of the surface wax is substantially higher in paraffin content than 40-50%.

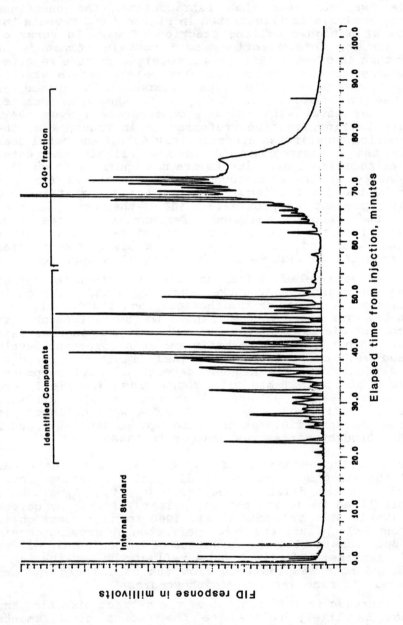

FIGURE 4 -- GC/FID Chromatogram of Sample #2.

TABLE 9 — Changes in Johnsongrass Surface Chemistry With Age and Drought Stress

COMPOUND TYPE	WT% OF IDENTIFIED COMPOUNDS							
	WITHOUT SEED HEADS NO. 3-89	WITH SEED HEADS NO. 4-89	WITH SEED HEADS NO. 2-88	DROUGHT STRESS NO. 5-88	GREENHOUSE 1-MONTH NO. 9-89	WITHOUT SEED HEADS NO. 6 -90	WITHOUT SEED HEADS NO. 8 -90	WITH SEED HEADS NO. 7-90
N-PARAFFINS	24	49	27	40	33	33	28	41
ALCOHOLS	62	17	11	11	47	44	51	26
ACIDS	14	34	63	49	20	23	21	33
CARBON NO.	WT% OF C_{28}-C_{32} ALCOHOLS							
28	5	16	18	21	20	3	9	5
30	17	35	41	38	22	16	18	20
32	78	49	41	41	58	81	73	75

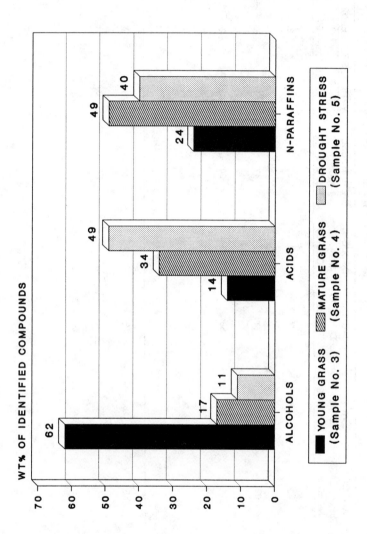

FIGURE 5 -- Change in Johnsongrass Surface Chemistry with Age and Drought Stress

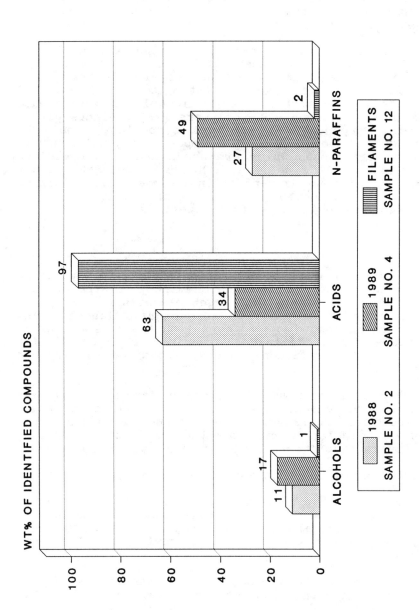

FIGURE 6 -- Change in Johnsongrass Surface Chemistry with Drought Stress in 1988 Compared to Same Field and Same Age in 1989

A loss of the filaments (acid component) from the surface has been observed in the field after heavy rainfall. The observed declining trend in polarity of johnsongrass surface from 24.4% to 16.5% from August to October may be related to the decline in the fatty acid constituents at the surface. Experiments are now planned to determine the polarity of johnsongrass surfaces after different rainfall simulations.

The chemical composition of the surface waxes is important in determining wettability of a leaf surface. The higher γ_c value for johnsongrass as compared to the γ_c wax value (28 vs 21) is likely related to -C-OH and C=O constituents at the surface. Both alcohol and acid groups increase the γ_c value on $-CH_2-$ polymers [26]. Plants with a higher acid component, at constant morphology, would be easier to wet. However, the acid component may be mainly in the form of trichromes which are above the actual surface layer. Thus, alcohols decline with age or drought, with a corresponding increase in n-paraffins. Under this condition of high n-paraffin content at the surface, nonaqueous lipid adjuvants would provide good foliar wetting.

CONCLUSIONS

Laboratory measurements of wetting (droplet movement and contact angle) as well as surface chemistry related properties of a series of mineral and vegetable oils were compared to measurements of coverage on johnsongrass leaves. The level of data collected was to establish trends in order to select tools for preliminary screening of adjuvants for applications or formulation reasons. Findings to date are summarized below:

1. Droplet movement down an inclined surface would not rank aqueous and nonaqueous materials for surface coverage. It would rank surface coverage for nonaqueous adjuvants. However, droplet movement had the smallest relationship to coverage as compared to the other wetting methods.

2. Volatility of the adjuvant is a factor.

3. The cosine of the contact angle on both wax or johnsongrass surface was a linear model with liquid surface tension. Critical surface tension for johnsongrass was 28, while the critical surface tension for the $C_{24}-C_{30}$ wax was 21.

4. Contact angles on $C_{24}-C_{30}$ wax and johnsongrass had a strong relationship including a linear model where θjg = 0.750 θwax + -7.94 for nonaqueous, r^2 = 0.819 and θjg = 1.101 θwax - 17.34, r^2 =0.876, when aqueous was included.

5. The linear models between θwax or $\theta johnsongrass$ and coverage were very similar. Sample preparation is

considered simpler for obtaining θwax measurements. θwax provided the highest relationships to coverage for ranking coverage on johnsongrass as a single parameter.

6. The surface wax of johnsongrass leaf blades changes with age and likely changes under severe drought conditions.

7. Approximately 60% of the chloroform extract remains to be identified and is believed to represent to a large degree esters. Experiments are now underway to confirm this hypothesis.

8. Odd numbered C_{27} to C_{33} normal paraffins are a major component on leaf blades. Specialized structures at the surface of a leaf (fatty acids) also play a part in the total surface chemistry, but the underlying epicuticular surface is predominantly alcohols for young plants and normal paraffins for mature plants.

REFERENCES

[1] McWhorter, C. G. and Barrentine, W. L., "Spread of Paraffinic Oil on Leaf Surfaces of Johnsongrass (*Sorghum halepense*)," Weed Science 36: 111-117, 1988.

[2] Barrentine, J. L. and Warren, G. F., "Isoparaffinic Oil as a Carrier for Chlorpropham and Terbacil," Weed Science, 18: 365-372, 1970.

[3] Chambers, G. V., "The Role of Orchex® 796 in Pesticide Applications," Proceedings of the International Conference on Pesticides in Tropical Agriculture, MARDI, Kuala Lumpur, 1987.

[4] Treacy, M. F., Benedict, J. H., Schmidt, K. M. and Anderson, R. M., "Mineral Oil: Enhancement of Field Efficacy of A Pyrethroid Insecticide Against the Boll Weevil (*Coleoptera: curculionidae*)," 1988 proceedings Beltwide Cotton Production Research Conference, National Cotton Council of America.

[5] Saunders, R. K. and Lonnecker, W. M., "Physiological Aspects of using Nonphytoxic Oils with Herbicides," Proceedings of the 21st North Central Weed Control Conference, Dec. 5-7, 1967: 62-63.

[6] Stevens, P. J. G. and Baker, E. A., "Factors Affecting the Foliar Absorption and Redistribution of Pesticides. I. Properties of Leaf Surfaces and Their Interactions with Spray Droplets," Pesticide Science, 19: 265-281, 1987.

[7] Wiese, A. F., "Oiling Herbicides," Proceedings of 1967 Meeting of National Petroleum Refiners Association, Washington, D.C., 1-7, 1967.

[8] Bandeen, J. D., "Wetting Agents, Oils, Are They A Plus for Atrazine," Crops and Soils, 21:15, 1969.

[9] Pesticides on Plant Surfaces, H. J. Cottrell, Ed., John Wiley and Sons, New York, 1987.

[10] McKay, B. M., "Surface Wetting-II," Pesticide Formulations and Application Systems: 10th Vol. ASTM STP 1078, 1990.

[11] Tulloch, A. P. and Hogge, J. R., "Gas Chromatographic-Mass Spectrometric Analysis of B-Diketone-Containing Plant Waxes," Journal of Chromatography, 157: 291-296, 1978.

[12] Van Valkenburg, J. W., "Terminology, Classification and Chemistry," 1-9, in Adjuvants for Herbicides, Weed Science Society of America, Champaign, Illinois, 1978.

[13] Wu S., Polymer Interface and Adhesion, Marcel Dekker, Inc., New York, 630, 1982. Chapter 5, "Surface Tension and Polarity of Solid Polymers."

[14] Fox, H. W. and Zisman, W. A., "The Spreading of Liquids on Low-energy Surfaces, III. Hydrocarbon Surfaces," Journal of Colloid Science, 7: 428-442, 1952.

[15] Barrentine, W. L. and McWhorter, C. G., "Johnsongrass (Sorghum halepense) Control with Herbicides in Oil Diluents," Weed Science, 36: 102-110, 1988.

[16] Hull, H. M., et al, "Action of Adjuvants on Plant Surfaces," 26-67, Adjuvants for Herbicides, Weed Science Society of America, Champaign, Illinois, 1978.

[17] Jeffree, C. E., "The Cuticle/Epicuticular Waxes and Trichomes of Plants, with Reference to Their Structure, Functions, and Evolution," 23-64, in Juniper and Southwood, Eds., Insects and the Plant Surface, Edward Arnold, London, 359, 1988.

[18] Tulloch, A. P. and Bergter, L., "Epicuticular Wax Composition of Echinochloa crusgalli, Phytochemistry, 19:145-146, 1980.

[19] Tulloch, A. P., et al, "A survey of Epicuticular Waxes Among Genera of Triticeae, 2. Chemistry," Canadian Journal of Botany 58: 2602-2615, 1980.

[20] Tulloch, A. P. and Weenink, R. O., "Composition of the Leaf Wax of Little Club Wheat," Canadian Journal of Chemistry, 47:3119-3126, 1969.

[21] Tulloch, A. P. and Hoffman, L. L., "Leaf Wax of Oats," Lipids, 8: 617-622, 1973.

[22] Tulloch, A. P. and Hoffman, L. L., "Epicuticular Waxes of Secale cereale and Triticale hexaploide Leaves," Phytochemistry, 13: 2535-2540, 1974.

[23] Bianchi, G., et al, "Composition and Structure of Maize Epicuticular Wax Esters," Phytochemistry, 28: 165-171, 1989.

[24] Kurtz, E. B., Jr., "The Relation of the Characteristics and Yield of Wax to Plant Age," Plant Physiology, 25: 269-278, 1950.

[25] McWhorter, C. G. and Paul, R. N., "The Involvement of Cork-Silica Cell Pairs in the Production of Wax Filaments in Johnsongrass (Sorghum halepense) Leaves, Weed Science, 37: 458-470, 1989.

[26] Adamson, A. W., Physical Chemistry of Surfaces, Interscience Publishers, London, 747, 1967.

Raj Prasad

SOME FACTORS AFFECTING HERBICIDAL ACTIVITY OF GLYPHOSATE IN RELATION
TO ADJUVANTS AND DROPLET SIZE

REFERENCE: Prasad, R., "Some Factors Affecting Herbicidal
Activity of Glyphosate in Relation to Adjuvants and Droplet
Size," Pesticide Formulations and Application Systems: 11th
Volume, ASTM STP 1112, Loren E. Bode and David G. Chasin, Eds.,
American Society for Testing and Materials, Philadelphia, 1992.

ABSTRACT: The influence of four adjuvants (Ethokem, Multifilm,
Regulaid and Tween-20) and four spray-droplet sizes (159, 332,
447, 575 µm) on efficacy and crop tolerance with a glyphosate
formulation were investigated for white birch [Betula papyrifera
L.] and white spruce [Picea glauca (Moench) Voss] under
greenhouse and laboratory conditions. It was found that some
adjuvants (Ethokem, and Tween-20) enhanced the effectiveness of
glyphosate sprays without damaging the crop (white spruce)
species. Tests with ^{14}C-glyphosate showed greater penetration
and translocation by birch leaves when an adjuvant was used. Of
the four droplet sizes tested, small droplets (159 µm) of
glyphosate were more phytotoxic than large drops (575 µm). The
implication of these findings in relation to herbicidal action of
glyphosate on forestry species is discussed.

KEYWORDS: application technology, forest weeds, Roundup,
surfactants, translocation.

It is a well established fact that herbicidal activity in plants
can be modified by many factors and conditions. Two factors in the
application scheme that may influence efficacy of herbicides are use
of adjuvants in the formulation and size of the spray droplet [1].

Spray adjuvants, by definition, are a class of chemicals designed
to modify and facilitate the effectiveness of the active ingredient
[2]. Although considerable research has been carried out on
herbicide and adjuvant interactions with weeds in agricultural crops
[3,4] very little information is available concerning the influence
of surfactants on forest weeds [5,6]. Three reasons are cited: (i)
the agricultural market is lucrative to manufacturers of these
products, (ii) forest weeds are perennial and hence more difficult to

Dr. Raj Prasad is a Research Scientist at Forest Pest Management
Institute, P.O. Box 490, Sault Ste. Marie, Ont. His current address:
Pacific Forestry Centre, 506 W. Burnside Road, Victoria, B.C. V8Z 1M5.

control and (iii) use of chemicals (herbicides and adjuvants) for weed
management in forestry is controversial and poses special economic and
environmental problems [7].

While application parameters such as droplet size, carrier volume
and herbicide concentration have been reported to affect the efficacy
of the active materials applied [1], there are conflicting views
regarding the role of droplet size in this respect. Some
investigators [8,9,10,11,12] claim that small droplets are more
efficaceous than large droplets, primarily because they provide better
coverage and demonstrate greater penetration and translocation than
larger ones.   Others [13,14,15,16,17] believe that smaller droplets
are not desirable because they are prone to drift and contribute to
contamination of the off-target areas. A third school of thought
[18,19] suggests that droplet size has no significant influence on
weed control under field conditions.

Herbicides are regarded as one of the most cost-effective tools in
the regeneration of forests, but glyphosate and other promising new
forest herbicides like hexazinone and triclopyr are much more
expensive than the traditionally-used 2,4-D [7]. Consequently there
is a need for research that will determine ways to improve the
efficacy of these herbicides under forestry conditions and reduce the
cost of their application. The objectives of this research were to
examine the influences of adjuvants and spray-droplet sizes on
glyphosate efficacy and crop tolerance. The present report describes
the effects of four surfactants (Ethokem, Multifilm, Regulaid and
Tween-20) and four droplet sizes (159, 332, 447, 575 μm) of glyphosate
on phytotoxicity to white birch seedlings.

MATERIALS AND METHODS

Plant Culture. White birch (Betula papyrifera L.) seeds were
germinated in sterilized soil kept at 20°C. When one month old, the
seedlings were transplanted into individual pots (15 x 15 x 10 cm)
filled with sterilized soil and cultivated in a greenhouse under
controlled conditions [temp. 20 $\pm$ 1°C; light-regime 2000 lux with a
16:8 h (light/darkness); and a relative humidity (RH) of 70 $\pm$ 15%].
Seedlings at the 6-8 leaf stage (4 months old) were used as test
plants.

For testing crop tolerance, 4-year-old white spruce seedlings
Picea glauca (Moench) Voss were removed from cold storage and induced
to flush under greenhouse conditions (similar to above) for a
considerable period (16 weeks) so that needles were completely
"hardened" prior to treatment.

Droplet Production. A specially designed droplet generator (Fig.
1) was used to produce monosized droplets of the four diameters to be
tested (159, 332, 447, and 575 μm). Built on principles developed by
Rayner and Haliburton [20], the device involves a low, but constant
flow of solution that is forced through a needle (ID:0.0004 mm) to
form a pendant drop. A spinning disc with attached blade, rotates
vertically at a constant speed and excises a single droplet from the
pendant drop at each revolution. When equilibrium is achieved

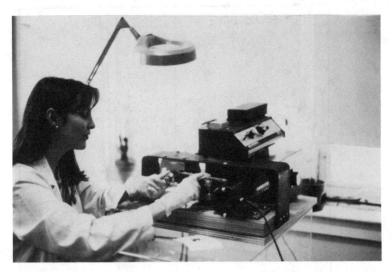

FIG. 1 -- Monosize droplet generator used for production of
homogeneous and uniform droplets.

between flow to the pendant drop and the rate of droplet excision, the
droplets become monosized. Monosized droplets of different diameters
are obtained by either varying the flow rate and disc's rotational
speed or the size of the attached blade.

To confirm that the aerodynamic droplets were indeed of known
diameter, monosized and numerically matched with the known RPM of the
disc, a dye was added to the test solution and the drops were caught
on Kromekote cards (Fig. 2). The resultant stains were counted and
measured. Stains from a range of aerodynamic droplets of known
diameter used in this study showed a linear relationship between
stains and droplets (Fig. 3).

Treatment of Plants. (i) With adjuvants: Preliminary bioassays
on standard greenhouse-grown seedlings determined that at 10% of the
maximum field dosage rate for glyphosate (2.1 kg AI/ha) the average
foliar damage to white birch was about 30%. This dosage level (0.21
kg AI/ha) was used for succeeding trials so that any enhancement in
phytotoxicity due to addition of adjuvants could be determined
relative to treatment with glyphosate alone. Glyphosate formulations
were prepared with Multifilm[1], Regulaid[2], and Tween-20[3] at 0.1% (v/v)
and Ethokem[4] at 0.5% (v/v). Some properties of these compounds are

---

[1] Multifilm is a non-ionic spreader-sticker of Celloidal Products
Ltd. Calif.

[2] Regulaid is a non-ionic activator-spreader of Kalo Agric.
Chemicals Inc. Kansas.

[3] Tween-20 is a non-ionic wetting agent of Atkemix Inc. Ont.

[4] Ethokem is a cationic wetting-spreading agent of Midkem Ltd. U.K.

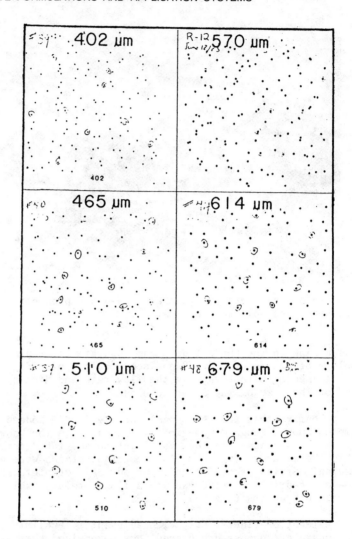

FIG. 2 -- Stains from small, medium and large homogeneous and uniform droplets produced by the monosize droplet generator. A 0.1% rhodamine dye was added to the glyphosate formulation to stain the imprints of droplets on Kromekote cards.

mentioned in the footnotes. Glyphosate and glyphosate + adjuvant formulations were sprayed onto foliage of test species in a spraying chamber. The conditions of spraying were: sprayer speed - 4 km/h; pressure - 206 kPa; nozzle-hydraulic flat fan-8005; and volume rate - 80 L/ha. Seedlings of both white birch and white spruce were also treated with a range of concentrations (0.1, 0.5, 1, 2% v/v) of each adjuvant alone to confirm that they themselves were not phytotoxic. After spraying, the plants were brought to a post-treatment chamber and maintained there under the same environmental conditions at which they were grown to allow symptoms of toxicity to develop.

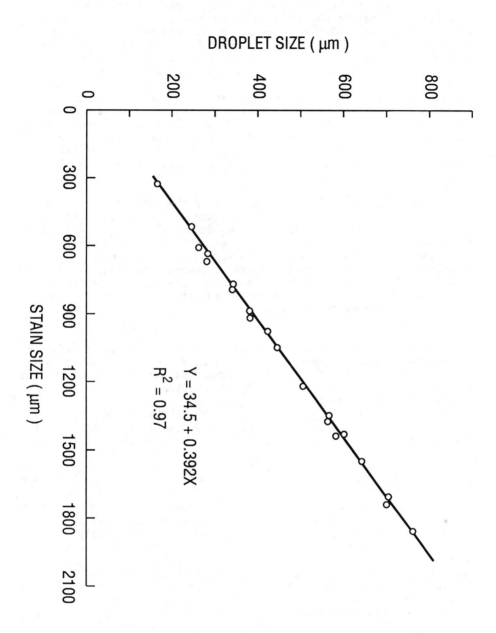

FIG. 3 -- Relationship between stain sizes of glyphosate and their corresponding drop sizes (a 0.1% rhodamine dye was added to the herbicide formulation to calculate the spread factor).

(ii) With different droplet sizes: To ensure a targeted application of herbicide to leaves or needles of the desired number and location, the seedlings were each covered with a plastic bag (Saran Wrap, Union Carbide, Toronto) and the target foliage exposed through a cut opening. For most treatments only the third fully-expanded leaf was exposed in this way. Treatment of individual plants was conducted in an enclosed plexiglass chamber to minimize inflight evaporation of the droplets, and to allow droplets emitted by the droplet generator to impact accurately on the target leaf. Volume and dosage rates were held constant (2 µL @ 1.0 kg AI/80L water/ha) for each application. By adding a tracer dye (Rhodamine 'B' dye 0.1% v/v) to the glyphosate formulation, droplets were clearly visible for counting and assessment of accurate distribution across the leaf lamina.

Response Measurement. Phytotoxicity in the adjuvant trials was assessed using two methods: i) scoring the foliar damage to each seedling on a percentage scale (0% = no damage; 100% = complete kill) [10], and ii) comparing the various treatments by changes in mean fresh weights of the plants over the period of symptom development. Time to response (necrosis, bleaching, epinasty, buckling, twisting, etc.) varied for the different glyphosate + adjuvant treatments, but 6 weeks was generally found to be sufficient time for these responses to stabilize. Assessments were conducted three and six weeks after application.

In the case of droplet-size effects, phytotoxicity was assessed by scoring the damage to the treated leaf (TL) and untreated parts (UP) using the above scale (0 to 100%). Translocation of the herbicide was grossly measured by the degree of damage or symptoms seen in the actively growing untreated parts (young leaves, shoots, apical buds). For more precise assay of translocation, ^{14}C-glyphosate was fed to a birch leaf and the course of absorption and translocation measured by a standard procedure.

Procurement of Chemicals

Adjuvants were obtained from different manufacturers whose details are given earlier. G3780A is a surfactant and was supplied by Monsanto Co. for studies with [^{14}C]-glyphosate. Roundup (glyphosate) was purchased from Monsanto Chemical Companies and was the standard commercial formulations recommended for weed control in coniferous forests. ^{14}C-glyphosate (sp. act., 9.3 mCi/mM) was obtained from Monsanto Chemical Company, St. Louis, Mo. as parent acid and converted to monoisopropyline salt by addition of isopropylamine in a 1:1 molecular ratio. Scintillation cocktail chemicals were purchased from Canlab Ltd. (Toronto) and other analytical reagents used for extraction of radio-activity were of pure quality.

Uptake of ^{14}C-glyphosate

The impact of an adjuvant on the degree of adsorption and translocation of herbicide was investigated using radio-labelled glyphosate. The third fully-expanded leaf from the apex of each white birch seedling was treated with ^{14}C-glyphosate either alone or in combination with surfactant (G3780A-0.05%)[5]. The procedure as described by Prasad et al. [3] involved pipetting a fixed dosage of active ingredient at 1200 ppm into a ring on the leaf lamina. The leaf was allowed to absorb the herbicide for two weeks in a growth chamber set at constant temperature (22 $\pm$ 1°C); light/dark (18/6 h); light intensity (17600 lux) and relative humidity (70 $\pm$ 15%). The ^{14}C-glyphosate was then extracted using the method for woody plants described by Sprankle et al. [22]. Radioactivity was measured by gross autoradiography [23] and by the standard scintillation counting technique [24].

Experimental Design and Statistical Treatment

All experiments used 15 replicates in each treatment and were repeated. Due to the variation both in seedling growth characteristics and the methods of treatment, analysis of variance was used to determine differences between herbicide-adjuvant mixes. Where necessary, data on percentages were transformed to angular arcsine to minimize error in the treatment [25]. In some cases, Duncan's Multiple Ranges Test (DMRT) [26] and regression analysis were employed to discern differences between treatments.

RESULTS AND DISCUSSION

Effect of Adjuvants. Assays conducted using recommended rates of each adjuvant mixed only in water revealed no phytotoxic properties on either crop or brush species. In the trials using the various adjuvants in a glyphosate formulation, (Table 1), the adjuvant Multifilm did not appear to modify the effectiveness of glyphosate whereas Ethokem, Regulaid and Tween-20 all markedly enhanced the phytotoxicity symptoms exhibited by the white birch seedlings. Ethokem produced the highest degree of this enhancement, followed by Tween-20 and Regulaid. These results are similar to those reported by other investigators [27,3,6,28]. This differential effect of Multifilm might be related to its sticking properties whereby glyphosate was adsorbed on its surface.

Adjuvant effects on uptake and translocation of ^{14}C-glyphosate. There are several ways by which adjuvants enhance phytotoxicity of herbicide formulations and these were described by Hodgson [2]. The theory that adjuvants do promote uptake and translocation of the herbicides is borne out in the ^{14}C-glyphosate trial (Table 2). Penetration and translocation to root and shoot was 40% greater in plants treated with the glyphosate + adjuvant formulation than in those treated with glyphosate alone. This increased mobility of herbicide into the root zone was shown in the autoradiograms as well.

Because glyphosate moves slowly in perennial forest species [6], the intensity of accumulation in roots after 2 weeks' exposure was

---

[5] G3780A is a non-ionic wetting agent supplied by Monsanto Co., St. Louis, Mo.

TABLE 1 -- Effects of glyphosate at 0.21 kg/ha and adjuvant combinations on changes in fresh weight and phytotoxicity of white birch leaves.

Treatment	Fresh Wt. (g/pot)	Phytotoxicity (%)
Control	16.3	0.0
Glyphosate	14.5	24.3
Glyphosate + Ethokem (0.5%)	10.4**	52.4***
Glyphosate + Multifilm (0.1%)	13.6	26.2
Glyphosate + Regulaid (0.1%)	11.9**	38.1**
Glyphosate + Tween-20 (0.1%)	12.1**	51.1**

*** Tukey's test, significant P = 0.01
** Tukey's test, significant P = 0.05

TABLE 2 -- Effects of an adjuvant (G3780A; 0.05%) on foliar penetration and translocation of ^{14}C-glyphosate (1200 ppm) in white birch harvested two weeks after treatment.

Treatment	Radioactivity (cpm/g fresh weight)				
	Leaves (L)	Stem (S)	Roots (R)	Total activity	Translocation (S+R)
White birch					
^{14}C-glyphosate	1617	50	375	2002(100%)	385(100%)
^{14}C-glyphosate + G3780A	2008	167	450	2625(131%)	617(143%)

still relatively low. With time, most of the herbicide eventually moved into the root zone [21,22,27,4,6] and accumulated there in sufficient quantity to prevent resprouting of the perennial weeds. Metabolism of glyphosate in roots is not common [22,27,29] and consequently translocation is a key factor in the control of resprouts. Even though adequate amount of adjuvants are present in the commercial formulation of glyphosate, the present study conducted under greenhouse conditions, demonstrates that efficacy at low dosages can be significantly augmented by the addition of some adjuvants. In a preliminary field trial with six different adjuvants it was found that addition of adjuvants to Roundup at high volume rates (300 L/ha) enhanced the control of aspen in small field plots. Thus it is possible that at very high volume rates of application of the commercial product, the innate concentration on the adjuvants get diluted and that addition of other surfactants enhances the efficacy of the herbicide. If such a trend is confirmed, it would suggest that the addition of surfactants could reduce the amount of active ingredient required, and thereby reduce the cost of application in forestry scenarios.

Effect of Droplet Sizes. Data relating to the phytotoxic response of white birch after treatment with four droplet sizes of glyphosate (Table 3) showed somewhat inverse relationship between droplet size and the phytotoxic response in the treated leaves i.e. small droplets were generally more phytotoxic than larger ones. Small droplets also on the whole, caused more damage, to the untreated part of the plant (U.P.) than larger droplets, presumably as a result of increased translocation of glyphosate. Neither small nor large droplets were found to be toxic to hardened spruce needles grown under similar greenhouse conditions. These results add support to the hypothesis that a glyphosate formulation applied in small droplets (159 to 332 μm) is generally more phytotoxic to broad-leaf plants than if applied in large droplets (575 μm). Similar findings have been reported by Buhler & Burnside [30] for other weeds. These observations imply that thorough coverage of weeds is necessary if optimum phytotoxicity is to be achieved. Field research is required to corroborate these findings and to better elucidate the limits of small droplets.

TABLE 3 -- Effects of droplet size of glyphosate on toxicity to white birch.

Drop size (μm)	No. of drops/ leaf[1]	Phytotoxicity			
		3 weeks[2]		6 weeks[2]	
		TL	UP	TL[3]	UP[3]
		---------------- % ----------------			
159	967	75 a	13 a	89 a	23 a[4]
332	103	65 ab	8 ab	68 ab	22 a
447	43	67 ab	5 b	71 a	15 b
575	21	52 b	6 b	60 b	15 b

[1] One leaf treated using 2 μL. Concentration 2.0 kg/80 L water.

[2] Each treatment analyzed separately.

[3] UP-TL-Untreated parts and treated leaf.

[4] Means for the same column followed by the same letter are not significantly different (P=0.05) DMRT.

CONCLUSION

Data presented in this study from a greenhouse experiment suggest that some factors such as addition of different adjuvants to glyphosate formulation as well changes in make-up of its droplets can modify the phytotoxicity of the active ingredient. If these findings can be duplicated under field conditions, then due consideration should be given to these factors while using herbicides on an operational scale. For example by employing appropriate adjuvant mixtures droplet sizes and volume rates, economy in the use of glyphosate can be achieved. This improved use, in turn, would be of considerable benefit to the clientele and the environment.

ACKNOWLEDGMENTS

The author is grateful to Dr. Leo Cadogan of this Institute for use of his monosize droplet generator and to Dal Travnick and Wanda Wright for technical assistance.

REFERENCES

[1]   Anderson, W. P., Weed Science: Principles 2nd ed. West Publishing Co. St. Paul, Minn., 1983.

[2]   Hodgson, R. H. (ed.), Adjuvant for Herbicides. A Monograph Published by the Weed Sci. Soc. Am., Champaign, IL., 1983, 143 pp.

[3]   Prasad, R., Foy, C. L. and Crafts, A. S., Effects of Relative Humidity on Absorption and Translocation of Dalapon. Weed Sci. 1967, 15:149-156.

[4]   McWhorter, G. G., The Use of Adjuvants. In "Adjuvant for Herbicides." WSSA monograph, 1983, 143 pp.

[5]   Prasad, R., Role of Some Adjuvants in Enhancing the Efficacy of Forest Herbicides. In "Adjuvants in Agrichemicals" Brandon, Man. Canada Edit P. Chow & C. Grant CRC Publn. Ohio, 1988, 1:159-165.

[6]   Prasad, R., Translocation of Glyphosate-C^{14} in Forest Weeds. Plant Physiol. 1984, 77(44): 152.

[7]   Sundaram, K. M. S. and Prasad, R., Research on Forest Herbicides in Canada: Problems, Progress and Status. In: "The 11th Workshop on Chemistry and Biochemistry of Herbicides." P.N.P. Chow and C.A. Grant (eds.), 1984, 45-52.

[8]   Ennis, W. B. and Williamson, R. E., The Influence of Droplet Size on Effectiveness of Low-volume Herbicide Sprays. Weeds, 1963, 11:67-72.

[9]   McKinlay, K. S., Brandt, S. A., Morse, P. and Ashford, R., Droplet Size and Phytotoxicity of Herbicides. Weed Sci. 1972, 20:450-452.

[10] Prasad, R., Droplet Size and Efficacy of Some Forest Herbicides. Res. Rept. Expt. Comm. Weeds, East Canada: 1985, p. 616.

[11] Rogers, R. B. and Kirkland, K., Billion Drop Technology Project-year 1. Trans. Am. Soc. Agric. Eng. 1985, 85:1631.

[12] Ambach, R. M. and Ashford, R., Effects of Variation in Drop Make-up on the Phytotoxicity of Glyphosate. Weed Sci., 1987, 30:221-224.

[13] Smith, M. H., Quantitative Aspects of Aqueous Spray Applications of 2,4-D Acid for Herbicide Purposes. 1946, Bot. Gaz.: 107, 546,

[14] Barzee, M.A. and Stroube, E.W., Low Volume Application of Pre-emergence Herbicide. Weed Sci. 1972, 2:176-180.

[15] Maybank, J., An Assessment of the Efficiency and Effectiveness of Herbicidal Applications. Sask. Res. Council Publication 1981, 799-2-D. 10 pp.

[16] Barry, J. W., Drift Predictions of Three Herbicide Tank-mixes. USDA-F.S. Davis Calif. Publn., 1985, 85:1-4.

[17] Grover, R., Schewchuk, S. R., Cessna, A. J., Smith, A. E., Hunter, J. H., Fate of 2,4-D Isooctyl Ester After Application to a Wheat Field. J. Environ. Qual., 1985, 14(2):203-210.

[18] Mullison, W. R., The Effect of Droplet Size Upon Herbicidal Effectiveness of Plant Growth Regulators. Down to Earth 1953, 9:11-13.

[19] Ayres, P., Taylor, W. A. and Cotterill, E. G., The Influence of Cereal Canopy and Application Method in Spray Deposition and Biological Activity of a Herbicide for Broad-leaved Weed Control. Crop Protection, 1984, 4:241-250.

[20] Rayner, A. C. and Haliburton, W., Rotary Device for Producing a Stream of Uniform Drops. Rev. Scientific Instr., 1955, 26(12):1124-1127.

[21] Sprankle, P., Meggitt, W. F. and Penner, D., Rapid Inactivation of Glyphosate in Soil. Weed Sci. 1975a, 23:224-228.

[22] Sprankle, P., Meggitt, W. F. and Penner, D. Absorption, Action and Translocation of Glyphosate. Weed Sci. 1975b, 23:235-240.

[23] Crafts, A. S. and Yamaguchi, S., The Autoradiography of Plant Materials. Agric. Stn. and Ext. Serv. Manual 35, Calif. Univ., 1964, 143 pp.

[24] Chase, G. F. and Rabinowitz, J. L., Principles of Radioisotope Methodology. Burgess Publishing Co., Minneapolis, U.S.A. 1964, 372 pp.

[25] Snedecor, G. W., Statistical Methods. Iowa State College Press, 1957, 534 pp.

[26] Duncan, D. B., Multiple Range and Multiple F Tests. Biometrics, 1955, 11:1-42.

[27] Gottrup, I., Sullivan, P. A., Schraa, K. J. and Vandenborn, W. H., Uptake, Translocation, Metabolism and Selectivity of Glyphosate in Canada Thistle and Leafy Spruce. Weed Res. 1976, 16:197-201.

[28] Wills, G. D., Factors Affecting Toxicity and Translocation of Glyphosate in Cotton. Weed Sci. 1978, 26:509-513.

[29] Wyrill, J. B. and Burnside, O. C., Absorption, Translocation and Metabolism of 2,4-D and Glyphosate in Common Milkweed and Hemp Dogbane. Weed Sci. 1976, 24:557-566.

[30] Buhler, D. D. and Burnside, O. C., Effects of Spray Components on Glyphosate Toxicity to Annual Grasses. Weed Sci. 1983, 31:124-130.

Frank A. Manthey, Richard D. Horsley, and John D. Nalewaja

RELATIONSHIP BETWEEN SURFACTANT CHARACTERISTICS AND THE PHYTOTOXICITY
OF CGA-136872.

REFERENCE: Manthey, F. A., Horsley, R. D., and Nalewaja,
J. D., "Relationship Between Surfactant Characteristics
and the Phytotoxicity of CGA-136872," Pesticide Formulations and Application Systems: 11th Volume, ASTM STP 1112,
Loren E. Bode and David G. Chasin, Eds., American Society
for Testing and Materials, Philadelphia, 1992.

ABSTRACT: Research was conducted to determine the relationship
between the phytotoxicity of CGA-136872 and the chemical and
physical properties of surfactants. All surfactants tested
greatly enhanced CGA-136872 phytotoxicity to quackgrass.
Surfactant enhancement of CGA-136872 phytotoxicity increased,
peaked, and decreased as surfactant hydrophilic:lipophilic
balance (HLB) increased. The peak or optimum HLB ranged from
11.9 to 14.3. Surfactants that differed in chemistry but had an
HLB in the optimum range differed in their enhancement of CGA-
136872 phytotoxicity to quackgrass. Surfactant characteristics
that promoted quackgrass control with CGA-136872 were low
molecular weight, low surface tension, low cloud point
temperature, low pour point temperature, and HLB between 11.9
and 14.3.

KEYWORDS: Beacon, quackgrass, HLB, molecular weight, surface
tension, cloud point temperature, pour point temperature

CGA-136872, trade name Beacon, is a postemergence herbicide to
control grass weeds in corn. Beacon is formulated as a 75% ai
dispersable granule. Spray adjuvants are essential to Beacon efficacy
[1]. Dry formulated postemergence herbicides generally require a
surfactant or oil spray adjuvant for efficacy [2].

Surfactants are used in pesticide formulations and in spray
application. Surfactants reduce surface tension of spray droplets
which usually results in increased droplet retention and spread on the
leaf surface [3]. Surfactant enhancement of herbicidal activity
generally takes place at much greater concentrations than that needed
for wetting. Thus, surfactants probably have more than one function
in herbicide enhancement. Surfactants have been reported to
solubilize cuticular leaf waxes [4], act as herbicide co-solvents [5,

Drs. Manthey, Horsley, and Nalewaja are Research Scientist,
Assistant Professor, and Professor, respectively, Department of Crop
and Weed Sciences, North Dakota State University, Fargo, ND 58105.

6] and increase apparent herbicide solubility in water [7].

Surfactants have hydrophilic and lipophilic properties. The hydrophilic:lipophilic balance (HLB) has been used to classify surfactant function in pesticide absorption [3]. Foliar absorption of polar compounds has been positively correlated to the HLB of the surfactant; while foliar absorption of nonpolar compounds has been negatively correlated to the HLB of the surfactant [5].

The chemical type of the surfactant has been suggested to be more important than its HLB [8]. The chemistry of the lipophilic portion of the surfactant molecule may be more important than the hydrophilic portion of the surfactant, particularly with herbicides having low water solubility [8, 9]. Thus, water soluble (polar) herbicides and oil soluble (nonpolar) herbicides would require different surfactants for maximum foliar absorption.

Surfactants have been reported to increase, have no effect, or decrease foliar absorption of active ingredients [7, 10]. Establishing some criterion for selecting a surfactant that maximizes efficacy of the active ingredient would be beneficial. An effective surfactant present in the herbicide formulation could increase the spectrum of weed control and reduce the need for use of spray adjuvants selected by the farmer.

Little research has been published concerning the relationship between surfactant properties and herbicide efficacy. The objective of this research was to determine the relationship between surfactant properties and surfactant enhancement of Beacon phytotoxicity to quackgrass.

EXPERIMENTAL METHOD

Quackgrass rhizomes were dug in the fall of 1989 and in the spring of 1990. Rhizomes were stored in the dark at 2 C until used.

Six rhizome sections having two nodes each were planted 2.5 cm deep in 0.5 L plastic pots containing a greenhouse potting mix. Plants were watered as needed and fertilized with a water soluble fertilizer 7 and 35 days after treatment. Natural light was supplemented for a 16-h photoperiod. Newly emerged quackgrass shoots were cut to 7-cm when at the 2-leaf stage and allowed to regrow to the 2- to 3-leaf stage before treatment. Insects were controlled with diazinon granules.

Beacon was added to the spray carrier (Fargo municipal water, 56 ppm Ca/Mg) and the spray mixture was adjusted to pH 5, 7, or 9 by titrating with 0.5 M HCl and 0.5 M NaOH. The pH adjusted spray mixture was added to test tubes containing surfactants. The pH of the final spray mixture was determined but not adjusted unless noted. Experimental units were blocked based on the number and size of shoots. Experiments were conducted in a randomized complete block design with a factorial arrangement of spray carrier pH and surfactants. Each treatment was replicated four times and each experiment was repeated once. Data were subjected to an analysis of variance and treatment means were separated using Fisher's Protected LSD at the p<0.05 probability level.

Treatments were applied approximately 45 min after mixing using a moving nozzle pot sprayer that delivered 160 L/ha spray volume. The soil was covered with vermiculite before treatment. The vermiculite was removed immediately after treatment to reduce possible Beacon absorption from the soil. Shoot injury ratings were determined 7, 14, 21, and 28 days after treatment. Shoot fresh weight was determined 28 days after treatment and shoot regrowth fresh weight was determined 28 days after the initial harvest.

Experiments were conducted to determine surfactant properties that are important in the enhancement of quackgrass control with Beacon applied in various spray solution pH's. The surfactant properties evaluated were selected based on the ready availability of this information and on the published literature [3-9]. Surfactant properties are presented in Table 1. Nonoxynol, octoxynol, and block copolymer surfactants were evaluated with Beacon at 30 g/ha in a spray solution pH of 5, 7, and 9. Each surfactant series was a separate experiment. Liquid surfactants were applied at 0.25% v/v and solid surfactants were applied at 0.25% wt/v in the spray carrier.

Surfactant enhancement of Beacon toxicity to quackgrass increased, peaked, and decreased as HLB increased. Stepwise regression [11] was used to generate quadratic equations that described phytotoxicity as a function of HLB and HLB^2 for the nonoxynol and octoxynol surfactant series at each spray solution pH. The peak or optimum HLB was determined by taking the first derivative and setting the equation equal to zero.

Stepwise multiple regression was used to generate the simplest equation that described quackgrass injury and fresh weight reduction as a function of nonoxynol, octoxynol, and block copolymer surfactant properties (Table 1). Multiple regression equations selected had the greatest coefficient of deterination (R^2) in which all terms in the equation accounted for a significant portion (p<0.05) of the total variance. Linear equations were generated using HLB, molecular weight (MWT), surface tension (ST), cloud point temperature (CPT), wetting (WET), and pour point temperature (PPT). Quadratic equations were generated using HLB, HLB^2, MWT, MWT^2, ST, ST^2, CPT, CPT^2, WET, WET^2, PPT, and PPT^2. HLB was not included in the multiple regression using the block copolymer data because individual block copolymers have a wide HLB range (Table 1).

The optimum HLB range was used as an initial index for evaluating surfactants of different chemistries. Ten surfactants were selected which had HLB values of 13 to 13.6. Chemistry and HLB of these surfactants are presented in Table 2. Surfactants were applied with Beacon at 25 g/ha in a spray solution pH of 5, 7, and 9. A second experiment was conducted to compare the effectiveness of surfactants which differed in HLB and chemistry. Beacon was applied at 25 g/ha. The spray solution was adjusted to pH 7. The chemistry and HLB of these surfactants are presented in Table 3.

TABLE 1 -- List of surfactants and their properties.

Surfactant[a]	Mwt[b]	EO	HLB	Surface tension 25 C, dynes/cm	Cloud point 1% aqueous	Wetting, % soln for 25 sec	Pour point C
Nonoxynol							
Triton N42	396	4	9.1	0.1	...	Not soluble	...	-26
Triton N57	440	5	10	0.1	29	< 0 C	0.084%	-32
Triton N101	660	9.5	13.4	0.1	30	54 C	0.047%	4
Triton N150	880	15	15	0.1	35	95 C	0.1%	18
Octoxynol							
Triton X45	426	5	10.4	0.1	28	< 0 C	0.055%	...
Triton X100	642	9.5	13.5	0.1	30	65 C	0.048%	7
Triton X102	778	12.5	14.6	0.1	32	88 C	0.064%	16
Triton X165	910	16	15.8	0.1	35	>100 C	0.33%	13
Triton X405	1966	40	17.9	0.1	46	>100 C	5% >300 sec	-4

Surfactant[a]	Mwt[b]	EO	HLB	Surface tension 25 C, dynes/cm	Cloud point % aq		Draves wetting, 25C % aq	sec	Pour point C
Block copolymers									
Pluronic 10R5	1950	...	12-18	0.1%	51	1	69	0.1	>360	15
Pluronic 17R4	2650	...	7-12	0.1%	44	1	46	0.1	>360	18
Pluronic 25R2	3600	...	7-12	0.1%	41	1	40	0.1	100	25
Pluronic L62	2500	...	1-7	0.1%	43	1	32	0.1	78	-4
Pluronic L63	2650	...	7-12	0.1%	43	1	34	0.1	>360	10
Pluronic L64	2900	...	12-18	0.1%	43	1	58	0.1	>360	16
Pluronic P104	5900	...	12-18	0.1%	33	1	81	0.1	30	32
Pluronic P123	5750	...	7-12	0.1%	34	1	90	0.1	35	31
Tetronic 50R4[a]	3740	...	7-12	0.1%	46	1	59	0.1	>360	-4
Tetronic 50R8[a]	10200	...	12-18	0.1%	53	1	88	0.1	>360	38
Tetronic 90R4[a]	7240	...	1-7	0.1%	43	1	43	0.1	40	12

[a]Not all surfactants are listed in 40 CFR. Tetronic surfactants are not approved substances for application to growing plants.

[b]Mwt is molecular weight; EO is ethylene oxide content; HLB is hydrophilic/lipophilic balance, and aq is aqueous.

TABLE 2 -- Chemistry of surfactants in the optimum HLB range.

Surfactant[a]	Chemistry	HLB
Dow Corning 193	Silicone glycol copolymer	13.6
Igepal CO630	Nonylphenoxypoly(ethyleneoxy)ethanol	13.0
Mapeg 600 MO	PEG (600) monooleate	13.6
Mapeg 600 MOT	PEG (600) monotallate	13.5
Mapeg 600 MS	PEG (600) monostearate	13.6
Mazol 80 MGK	Ethoxylated mono- and diglycerides	13.1
Sterox NJ	Ethoxylated alkylphenol	13.0
Triton X100	Octyl phenoxy polyethoxy ethanol	13.5
Triton N101	Nonyl phenoxy polyethoxy ethanol	13.4
T-Det EPO P104	Block copolymer of propylene oxide and ethylene oxide	13.0

[a]Not all surfactants are listed in 40 CFR.

TABLE 3 -- Chemistry and HLB of surfactants used to determine the efficacy of surfactants in enhancing Beacon phytotoxicity to quackgrass.

Surfactant[a]	Chemistry and physical state	HLB
Atlox 3403F	Polyoxyethylene ether polyoxyethylene glyceride alkyl aryl sulfonate blend	
	Liquid	14.0
Dow Corning 193	Silicone glycol copolymers.	
	Liquid	13.6
Emerest 2654	PEG 400 Monopelargonate nonionic	
	Liquid	14.3
Emulphor VN430	Ethoxylated fatty acid, polyoxyethylated (5) oleic acid	
	Liquid	7.7
Emulphor ON877	Polyoxyethylated (20) oleyl alcohol	
	Liquid	15.4
GAFAC RS710	Anionic free acid of a complex organic phosphate ester	
	Liquid	...
Igepal CO630	Nonyl phenoxypoly(ethylenoxy) ethanol	
	Liquid	13.0
Pluronic P104	Block copolymers of ethylene oxide and propylene oxide	
	Paste	12-18
Pluronic 25R2	Block copolymers of propylene oxide and ethylene oxide	
	Liquid	1-7
Renex 36	Polyoxyethylene (6) tridecyl ether	
	Liquid	11.4
Surfadone LP100	N-Octylpyrrolidone	
	Liquid	...
Tergitol 15-S-12	C11-C15 Secondary alcohol ethoxylate	
	Liquid	14.5
Tetronic 904	Tetra-functional block copolymers of ethylene oxide and propylene oxide	
	Paste	12-18
Tetronic 90R4	Tetra-functional block copolymers of propylene oxide and ethylene oxide	
	Liquid	1-7
T-Mulz W	Anionic/nonionic blend of calcium alkylaryl sulfonate with polyoxyethylene ethers	
	Liquid	14.4
Tween 85	Polyoxyethylene (20) sorbitan trioleate	
	Liquid	11.0
Witconate P1059	Amine alkylaryl sulfonate	
	Liquid	

[a]Not all surfactants are listed in 40 CFR.

RESULTS AND DISCUSSION

Surfactant enhancement of Beacon phytotoxicity increased, peaked, and decreased as HLB increased. The peak or optimum HLB values and the R^2 for the equations used to determine the optimum HLB values for quackgrass injury and shoot fresh weight reduction at each spray solution pH are presented in Table 4. The optimum HLB values ranged from 12.1 to 14.3 for the nonoxynol surfactants and from 11.9 to 13.7 for the octoxynol surfactants. Surfactants used to enhance pesticide activity generally have HLB values of 10 to 13 [3]. The optimum HLB values were greater at pH 5 than 9 with the nonoxynol surfactants but were lower at pH 5 than 9 with the octoxynol surfactants. Thus, pH affect on the optimum surfactant HLB for enhancement of Beacon phytotoxicity may depend on surfactant chemistry.

Beacon applied alone was more phytotoxic at pH 9 than 5 (Table 5). At pH 9, surfactants did not differ in their enhancement of quackgrass control with Beacon regardless of surfactant chemistry. The high Beacon activity at pH 9 probably masked any differences in level of surfactant enhancement. However, at pH 5 Mapeg 600 MS and Mazol 80 MGK were less effective than the other surfactants, as determined by shoot regrowth. Mazol 80 MGK was the least effective surfactant. Mazol 80 MGK dissolved poorly in the spray solution. These data indicate that the chemistry of a surfactant affects enhancement of Beacon phytotoxicity so that surfactants with similar HLB values can differ in enhancement of Beacon for quackgrass control.

TABLE 4 -- Optimal HLB and coefficient of determination (R^2) values for nonoxynol and octoxynol surfactant series for injury and shoot fresh weight reduction of quackgrass.

Series	Spray sol'n pH		Injury 21	Injury 28	Shoot fresh weight reduction 28
Nonoxynol	5	HLB	14.3	13.1	13.0
		R^2	0.81	0.97	0.92
	7	HLB	13.2	12.7	13.0
		R^2	0.82	0.99	0.91
	9	HLB	12.6	12.1	12.5
		R^2	0.53	0.81	0.70
Octoxynol	5	HLB	11.9	12.8	12.5
		R^2	0.78	0.93	0.76
	7	HLB	13.5	13.2	13.3
		R^2	0.82	0.94	0.80
	9	HLB	13.3	13.2	13.7
		R^2	0.69	0.93	0.80

The "Days after treatment" spans the Injury (21, 28) and Shoot fresh weight reduction (28) columns.

Mapeg 600 MO was more effective than Mapeg 600 MS in enhancing Beacon control of quackgrass shoots 28 days after treatment and subsequent regrowth (Table 5). These two surfactants have similar hydrophilic chemistry and HLB but differ in their lipophilic chemistry. The lipophilic chemistry apparently affected the ability of these surfactants to enhance Beacon phytotoxity to quackgrass. These surfactants would differ in their ability to solubilize Beacon and/or cuticular components. Lipophilic portion of a surfactant is very important in the solubilization of nonpolar compounds [6, 12]. These results agree with previous findings that the chemistry of the surfactant may be more important than HLB to the enhancement of herbicide activity [8].

TABLE 5 -- Quackgrass response to Beacon at 25 g/ha applied with surfactants that have HLB values in the optimum range.

Surfactant[a]	Spray sol'n pH	Days after treatment				
		14	21	28	28 Shoot fresh weight reduction	Regrowth fresh weight reduction
		Injury				
		----------------- % -----------------				
None	5	9	8	9	9	27
Dow Corning 193	5	90	94	97	94	99
Igepal CO630	5	91	96	98	96	97
Mapeg 600 MO	5	88	95	98	95	99
Mapeg 600 MOT	5	89	94	97	94	87
Mapeg 600 MS	5	85	90	91	91	67
Mazol 80 MGK	5	70	73	70	69	39
Sterox NJ	5	90	95	98	95	95
Triton X100	5	88	95	96	94	90
Triton N101	5	92	95	98	95	99
T-Det EPO P104	5	88	92	96	95	87
None	9	31	35	31	30	16
Dow Corning 193	9	91	96	98	95	98
Igepal CO630	9	93	96	98	96	100
Mapeg 600 MO	9	93	98	99	97	100
Mapeg 600 MOT	9	92	95	97	95	95
Mapeg 600 MS	9	92	96	98	95	95
Mazol 80 MGK	9	89	95	98	95	90
Sterox NJ	9	96	98	100	97	100
Triton X100	9	91	97	98	95	100
Triton N101	9	96	98	99	96	98
T-Det EPO P104	9	91	98	99	96	98
LSD(0.05)		7	6	7	8	20

[a]Not all surfactants are listed in 40 CFR.

Equations were not generated using stepwise multiple regression that had greater R^2 values than those for the quadratic equations using HLB and HLB^2 for the nonoxynol and octoxynol surfactants (Table 4). R^2 values generally were high for both the nonoxynol and octoxynol surfactant series. For example, depending on spray solution pH quadratic equations for injury 28 days after treatment accounted

Mapeg 600 MO was more effective than Mapeg 600 MS in enhancing Beacon control of quackgrass shoots 28 days after treatment and subsequent regrowth (Table 5). These two surfactants have similar hydrophilic chemistry and HLB but differ in their lipophilic chemistry. The lipophilic chemistry apparently affected the ability of these surfactants to enhance Beacon phytotoxity to quackgrass. These surfactants would differ in their ability to solubilize Beacon and/or cuticular components. Lipophilic portion of a surfactant is very important in the solubilization of nonpolar compounds [6, 12]. These results agree with previous findings that the chemistry of the surfactant may be more important than HLB to the enhancement of herbicide activity [8].

TABLE 5 -- Quackgrass response to Beacon at 25 g/ha applied with surfactants that have HLB values in the optimum range.

		Days after treatment				
		14	21	28	28	
Surfactant[a]	Spray sol'n pH	Injury			Shoot fresh weight reduction	Regrowth fresh weight reduction
		---------------- % --------------------				
None	5	9	8	9	9	27
Dow Corning 193	5	90	94	97	94	99
Igepal CO630	5	91	96	98	96	97
Mapeg 600 MO	5	88	95	98	95	99
Mapeg 600 MOT	5	89	94	97	94	87
Mapeg 600 MS	5	85	90	91	91	67
Mazol 80 MGK	5	70	73	70	69	39
Sterox NJ	5	90	95	98	95	95
Triton X100	5	88	95	96	94	90
Triton N101	5	92	95	98	95	99
T-Det EPO P104	5	88	92	96	95	87
None	9	31	35	31	30	16
Dow Corning 193	9	91	96	98	95	98
Igepal CO630	9	93	96	98	96	100
Mapeg 600 MO	9	93	98	99	97	100
Mapeg 600 MOT	9	92	95	97	95	95
Mapeg 600 MS	9	92	96	98	95	95
Mazol 80 MGK	9	89	95	98	95	90
Sterox NJ	9	96	98	100	97	100
Triton X100	9	91	97	98	95	100
Triton N101	9	96	98	99	96	98
T-Det EPO P104	9	91	98	99	96	98
LSD(0.05)		7	6	7	8	20

[a]Not all surfactants are listed in 40 CFR.

Equations were not generated using stepwise multiple regression that had greater R^2 values than those for the quadratic equations using HLB and HLB^2 for the nonoxynol and octoxynol surfactants (Table 4). R^2 values generally were high for both the nonoxynol and octoxynol surfactant series. For example, depending on spray solution pH quadratic equations for injury 28 days after treatment accounted

for 81 to 99% of the total variance using data from the nonoxynol
surfactant experiment and 93 to 94% of the total variance using the
data from the octoxynol surfactant experiment. This would indicate
that HLB could be used to predict surfactant effectiveness with
Beacon; however, surfactants not having an HLB between 11.9 and 14.3
were also quite effective (Table 6). For example, the HLB of Tetronic
90R4 (1 to 7) is below and the HLB of Emulphor ON877 (15.4) is above
the optimum HLB range (11.9 to 14.3) but both surfactants were very
effective with Beacon (Tables 3 and 6). Thus, the selection of
surfactants for Beacon requires consideration of characteristics in
addition to HLB. The optimum HLB range may be the same for
surfactants that differ in chemistry. However, certain surfactant
chemistries appear to be more effective than other chemistries in
enhancing Beacon phytotoxicity.

TABLE 6 -- Quackgrass response to Beacon at 25 g/ha applied in a
spray solution of pH 7 with surfactants that differ in
chemistry and HLB.

Surfactant[a]	HLB	Days after treatment			28 Shoot fresh weight reduction	Regrowth fresh weight reduction
		14	21	28		
		Injury			%	
None	...	5	3	9	12	10
Atlox 3403F	14.0	84	93	96	94	99
Dow Corning 193	13.6	91	98	99	97	100
Emerest 2654	14.3	90	96	98	96	100
Emulphor VN430	7.7	88	95	98	96	88
Emulphor ON877	15.4	90	97	99	97	100
Gafac RS710	...	88	95	98	96	100
Igepal CO630	13.0	88	96	97	96	99
Pluronic P104	12-1	87	94	97	95	100
Pluronic 25R2	1-7	90	96	98	96	96
Renex 36	11.4	91	98	99	96	100
Surfadone LP100	...	82	92	92	91	76
Tergitol 15-S-12	14.5	91	98	100	98	100
Tetronic 904	12-1	89	96	97	95	100
Tetronic 90R4	1-7	86	95	98	96	100
T-Mulz W	14.4	87	95	98	96	100
Tween 85	11.0	80	91	95	94	76
Witconate P1059	...	82	92	95	93	68
LSD(0.05)		6	4	4	4	16

[a]Not all surfactants are listed in 40 CFR.

Linear and quadratic multiple regression equations generated
using data from quackgrass response to Beacon applied with block
copolymers are presented in Table 7. R^2 values were not greater for
cubic equations than quadratic equations. R^2 values of the multiple
regression equations were similar for percent fresh weight reduction
and shoot injury 28 days after treatment. R^2 values were lower at pH

9 than 5.   The apparent differences between surfactants in their enhancement of Beacon toxicity to quackgrass were much less at pH 9 than 5 making it difficult to establish a relationship between surfactant properties and Beacon efficacy at pH 9.

The linear multiple regression equations consisted of molecular weight, surface tension, and cloud point temperature (Table 7).   The linear regression equations indicate that, depending on spray solution pH, Beacon phytotoxicity increased 1 to 4% with a 1000 decrease in molecular weight; and increased 1 to 2% with a 1 dyne/cm decrease in surface tension or a 10 C decrease in cloud point temperature.

TABLE 7 -- Equations[a] generated by stepwise regression using the block copolymer data[b].

Percent fresh weight reduction (PCFWTRED).

Al1 data.

$PCFWTRED = 163.01323 - 0.0026(MWT) - 1.4201(ST)_2 - 0.1333\ CPT$ $\quad R^2_2 = 0.61$
$PCFWTRED = 148.5409 + 0.5343(PPT) - 0.02706(PPT^2) - 1.3860(ST) - 0.00000011\ (MWT^2)$ $\quad R^2 = 0.68$

Spray solution pH = 5

$PCFWTRED = 172.7898 - 0.0030(MWT) - 1.5899(ST) - 0.1643(CPT)$ $\quad R^2_2 = 0.73$
$PCFWTRED = 155.2005 + 0.5564(PPT) - 0.02927(PPT^2) - 1.5381(ST) - 0.00000013\ (MWT^2)$ $\quad R^2 = 0.80$

Spray solution pH = 7

$PCFWTRED = 178.1308 - 0.0032(MWT) - 1.722(ST) - 0.1532(CPT)$ $\quad R^2_2 = 0.67$
$PCFWTRED = 324.8437 + 0.9089(PPT) - 0.05266(PPT^2) - 9.1647(ST) + 0.08466(ST^2)$ $\quad R^2 = 0.77$

Spray solution pH = 9

$PCFWTRED = 134.1245 - 0.0020(MWT) - 0.9181(ST)$ $\quad R^2_2 = 0.47$
$PCFWTRED = 130.7050 + 0.00621(PPT^2) - 0.9190(ST) - 0.00000010(MWT^2)$ $\quad R^2 = 0.52$

Above ground shoot injury 28 days after treatment (INJURY28).

Al1 data

$INJURY28 = 161.1209 - 0.0022(MWT) - 1.3097(ST) - 0.1248(CPT)$ $\quad R^2_2 = 0.57$
$INJURY28 = 148.5310 + 0.4370(PPT) - 0.0229(PPT^2) - 1.2851(ST) - 0.00000009(MWT)$ $\quad R^2 = 0.62$

Spray solution pH = 5

$INJURY28 = 177.2343 - 0.0032(MWT) - 1.5928(ST) - 0.1562(CPT)$ $\quad R^2_2 = 0.68$
$INJURY28 = 158.3694 + 0.5298(PPT) - 0.02750(PPT^2) - 1.5121(ST) - 0.00000016(MWT^2)$ $\quad R^2 = 0.74$

Spray solution pH = 7

$INJURY28 = 149.8384 - 0.0038(MWT) - 0.9417(ST) - 0.0280(WET1)$ $\quad R^2_2 = 0.66$
$INJURY28 = 58.3582 + 0.5285(PPT) - 0.01909(PPT^2) + 0.01485(MWT) - 0.00000149(MWT^2)$ $\quad R^2 = 0.71$

Spray solution pH = 9

$INJURY28 = 129.2047 - 0.0012(MWT) - 0.7543(ST)$ $\quad R^2_2 = 0.55$
$INJURY28 = 129.1774 + 0.2223(PPT) - 0.01273(PPT^2) - 0.8266(ST)$ $\quad R^2 = 0.60$

[a] MWT is molecular weight; ST is surface tension of 0.1% aqueous solution at 25 C; CPT is cloud point temperature; and PPT is pour point temperature.
[b] The number of observations for each regression equation was 88.

The quadratic multiple regression equations consisted of pour point temperature, surface tension, and molecular weight (Table 7). The quadratic multiple regression equations indicate that Beacon phytotoxicity increased as pour point temperature, surface tension, and molecular weight decreased.

Molecular weight may affect the rate of surfactant movement into and through the cuticle and/or cell membrane. The rate of diffusion through membranes is slower for high molecular weight compounds compared to low molecular weight compounds of similar polarity. The rate of membrane penetration would affect the rate of herbicide absorption particularly if the surfactant acts as a co-solvent with the herbicide which together penetrates the cuticle [5, 6].

Surface tension may relate to wetting or spreading across the leaf surface; although wetting per se did not account for a significant portion (p<0.05) of the total variance of either the linear or quadratic multiple regression equations. The actual area of the spray deposit does not always relate to the area of wetting, due to the receding contact angle of the evaporating droplet [5, 12]. Surfactants reduce surface tension of spray droplets which usually results in increased droplet retention and spread on the leaf surface. An increase in spray droplet retention will result in increased herbicide deposition on the leaf surface.

Cloud point temperature may relate to the solubilization of Beacon in the spray solution. Aqueous solutions of polyoxyethylenated nonionic surfactants become turbid on being heated to a temperature known as the cloud point, following which there is a phase separation of the solution into two phases. As the temperature increases, micellar growth and increased intermicellar attraction cause the formation of particles that are so large that the solution becomes visibly turbid [13]. Upon cooling the mixture, the two phases merge to form a clear solution. Nonpolar compounds may have increased solubility in surfactant micelles as the cloud point temperature of the surfactant is approached [13]. Conversely, solubilization of polar compounds generally decreases as temperature approaches the cloud point. Cloud point should be higher than temperatures encountered during herbicide application. If the cloud point temperature is too low the surfactant may phase out of solution which would greatly reduce surfactant effectiveness in enhancing herbicide phytotoxicity.

Beacon solubilization by a surfactant may be important to activity. Solubilization may protect Beacon from degradation that occurs at low spray solution pH [1]. Solubilization of Beacon may enhance its activity by increased foliar absorption and/or decreased degradation.

Beacon phytotoxicity generally increased as pour point temperature decreased. A surfactant that is liquid at room temperature has a lower pour point temperature than a surfactant that is solid at room temperature. These data indicate that liquid surfactants would be more effective than solid surfactants in enhancing Beacon phytotoxicity to quackgrass. The affect of pour point temperature on surfactant enhancement of Beacon phytotoxicity may relate to the physical state of the surfactant once the spray droplet has evaporated on the the leaf surface. Diffusion into the cuticle may be slower for a solid than for a liquid surfactant. The rate of cuticular penetration would affect the rate of herbicide absorption particularly if the herbicide moves with the surfactant through the cuticle [5, 6].

Quackgrass control was greater when Beacon was applied in a spray solution of pH 9 than 5. Surfactants enhanced Beacon control of quackgrass. Apparent differences between surfactants were greater when the surfactants were applied with Beacon in a spray solution of pH 5 than 9. However, less enhancement of quackgrass control was possible at pH 9 than 5 because Beacon phytotoxicity approached complete control at pH 9. Surfactant characteristics that promoted quackgrass control with Beacon were low molecular weight, low surface tension, low cloud point temperature, low pour point temperature, and HLB between 11.9 and 14.3.

ACKNOWLEDGEMENTS

The authors acknowledge the assistance of Patricia Evans and Genece Hanson.

REFERENCES

[1] Technical Release, Beacon Herbicide, Ciba-Geigy Corporation, Greensboro, North Carolina, 1989.
[2] Crop Protection Chemicals Reference, Chemical and Pharmaceutical Press and John Wiley and Sons, New York, 1988.
[3] McWhorter, C. G., "The Use of Adjuvants," in Adjuvants for Herbicides, Weed Science Society of America, Champaign, Illinois, 1982, pp. 10-25.
[4] Kuzych, I. J. and Meggitt, W. F., "Alteration of Epicuticular Wax Structure by Surfactants," Micron and Microscopica Acta Vol. 14, 1983, pp. 279-280.
[5] Stevens, P. J. G. and Bukovac, M. J., "Studies on Octylphenoxy Surfactants. Part 2: Effects on Foliar Uptake and Translocation," Pesticide Science, Vol. 20, 1987, pp. 37-52.
[6] Silcox, D. and Holloway, P. J., "Foliar Absorption of Some Nonionic Surfactants from Aqueous Solutions in the Absence and Presence of Pesticidal Active Ingredients," in Adjuvants and Agrochemicals Volume I, CRC Press, Boca Raton, Florida, 1989, pp. 115-128.
[7] Temple, R. E. and Hilton, H. W. "The Effect of Surfactants on the Water Solubility of Herbicides, and the Foliar Phytotoxicity of Surfactants," Weeds, Vol. 11, 1963, pp. 297-300.
[8] Jansen, L. L., "Enhancement of Herbicide Activity: Relation of Structure of Ethylene Oxide Ether-Type Nonionic Surfactants to Herbicidal Activity of Water-Soluble Herbicides," Journal of Agricultural and Food Chemistry, Vol. 12, 1964, pp. 223-227.

[9]   Tan, S. and Crabtree, G. D. Crabtree, "Relationship of Chemical
      Classification and Hydrophilic-Lipophilic Balance of Surfactants
      to Upper Leaf-Surface Penetration of Growth Regulators in
      Apples," presented at the Second International Symposium on
      Adjuvants for Agrichemicals, Blacksburg, Virginia, 1989, Abstr.
      No. 17.
[10] Smith, L. W. and Foy, C. L., Foy., "Interactions of Several
      Paraquat-Surfactant Mixtures, Weeds Vol. 15, 1967, pp. 67-72.
[11] SAS Users Guide, SAS Institute, Cary, North Carolina, 1988.
[12] Furmidge, C. G. L., Physico-Chemical Studies on Agricultural
      Sprays. I. General Principles of Incorporating Surface-Active
      Agents as Spray Supplements," Journal of Science Food Agriculture,
      Vol. 10, 1959, pp. 267-273.
[13] Rosen, M. J., "Solubilization by Solutions of Surfactants:
      Micellar Catalysis," in Surfactants and Interfacial Phenomena,
      John Wiley and Sons, New York, 1989, pp. 170-206.

Christopher M. Riley[1], Charles J. Wiesner[2], Donald W. Scott[3], Julie Weatherby[4] and Robert G. Downer[5]

EVALUATING THE FIELD EFFICACY OF *Bacillus thuringiensis* BERLINER AGAINST THE WESTERN SPRUCE BUDWORM (LEPIDOPTERA:TORTRICIDAE)

REFERENCE: Riley, C. M., Wiesner, C. J., Scott, D. W., Weatherby, J., and Downer, R. G., "Evaluating the Field Efficacy of *Bacillus thuringiensis* Berliner Against the Western Spruce Budworm (Lepidoptera:Tortricidae)," Pesticide Formulations and Application Systems: 11th Volume, ASTM STP 1112, Loren E. Bode and David G. Chasin, Eds., American Society for Testing and Materials, Philadelphia, 1992.

ABSTRACT: A detailed assessment of spray deposition and efficacy of Dipel 6AF and Thuricide 48LV was carried out as part of the 1988 western spruce budworm control program. An analysis of covariance model was used in an attempt to examine the relationship between deposit density and field efficacy (larval mortality and defoliation) for the two formulations. The model could not be justified and relationships between the variables were investigated using chi-aquare analysis.

KEYWORDS: *Bacillus thuringiensis*, efficacy, spray deposition, spruce budworm.

[1] Pesticide Application Specialist, Research and Productivity Council, 921 College Hill Road, Fredericton, New Brunswick, Canada, E3B 6Z9.

[2] Senior Scientist, Research and Productivity Council, 921 College Hill Road, Fredericton, New Brunswick, Canada, E3B 6Z9.

[3] Entomologist, U. S. Department of Agriculture, Forest Service, Wallowa-Whitman National Forest, Forest and Range Science Laboratory, 1401 Gekeler Lane, La Grande, Oregon, 97850.

[4] Forest Entomologist, U. S. Department of Agriculture, Forest Service, Forest Pest Management, Boise Field Office, 1750 Front Street, Boise, Idaho, 83702.

[5] Statistical Consultant, University of New Brunswick, P. O. Box 4400, Fredericton, New Brunswick, Canada E3B 5A3.

Field tests of aerially applied insecticides, particularly biological insecticides, are subject to a high degree of variability. Weather conditions during the spray application and in the post-spray period influence both the amount of spray reaching the larval microhabitat and the subsequent vulnerability of the target insect to deposited spray [1 - 3]. The precise height at which the spray is released from the aircraft, the type of aircraft used, the physical/chemical nature of the spray formulation and the characteristics of the target stand are also variables which profoundly influence the deposition of spray in the microhabitat of the target insect.

It is, therefore, not sufficient to compare spray applications on the basis of "application rate" alone. Numerous experiments have shown that two well-replicated spray applications can result in very different levels of spray deposit on the target foliage (Kettela and Wiesner, unpublished). Such experimental variability could be dealt with by conducting large numbers of replicates, thereby overcoming the intrinsic variability in deposit, or by investigating the relationship between biological effectiveness and actual spray deposit rather than with application rate [4].

By measuring spray deposit, larval mortality and defoliation on each sample tree in the treated area and using the appropriate terms in an analysis of covariance model it should be possible to examine the relationships between deposit density and biological effectiveness. The characteristics of the resulting relationships are then a function of the toxicity of the formulation, the half life of the formulation activity, post-spray weather and the vigour of the pest population.

This study was carried out in conjunction with the Meacham Pilot Project on the Umatilla National Forest near Pendleton, Oregon as part of the 1988 Western Spruce Budworm Control Program. The Meacham Pilot Project was designed to evaluate the operational use of two commercial water-based formulations of the microbial pesticides *Bacillus thuringiensis* Berliner (B.t.) against the western spruce budworm (*Choristoneura occidentalis*, Freeman).

Specific objectives of this study were:
(1)   to quantify B.t. spray deposition on midcrown Douglas-fir and grand fir foliage in each of six spray blocks (i.e. three spray blocks per treatment);
(2)   to quantify the biological effectiveness and relative efficacy of Dipel 6AF and Thuricide 48LV;
(3)   to investigate deposit/field efficacy relationships for Dipel 6AF and Thuricide 48LV.

[1]   Abbott Laboratories, North Chicago, Illinois
[2]   Sandoz Crop Protection Corp., Des Plaines, Illinois

MATERIALS AND METHODS
Formulation and Application Parameters
Spray applications were made by Hiller 12E Soloy helicopters equipped with six model 360A-1 Beecomist rotary atomizers which were calibrated to give a droplet spectrum with a nominal $D_{v0.5}$ of 137 to 150 μm. Each formulation was applied undiluted at a rate of 39 BIU/ha (3.1 L/ha). The aircraft flew at a nominal speed of 72 km/h with a flight line separation ("swath width") of 34 m and an emission rate of 2.8 L/min/atomizer. The two aqueous formulations were dyed with the non-insecticidal dye Erio Acid Red XB-400% (0.2% W/V) to permit visualization of spray deposition on foliage. Three replicate spray blocks were assigned to each treatment as well as three untreated control areas. Block size ranged from 2200 ha to 4500 ha[3].

Sampling Design and Procedures
Within each spray block, seventy-five sample trees arranged in a total of twenty-five plots, each containing three trees, and representative of the range of elevations and aspects within each block were selected. In mixed conifer stands, sample trees were selected from the predominant host species. Each plot consisted of three Douglas-fir, *Pseudotsuga menziesii* (Mirbel) Franco, or three grand fir, *Abies grandis* (Douglas) Lindley, trees which were approximately 6 m to 13 m tall with full crowns and mostly open grown.

Spray deposition from aerially applied ULV sprays varies greatly, not only between different trees but also between locations within any one sample tree [2] [4 - 9]. If relationships between spray deposit and biological efficacy are to be meaningfully investigated, it is essential that pre-spray budworm populations, spray deposition, post-spray budworm populations and defoliation be assessed on branches from essentially the same location within any one sample tree, i.e. immediately adjacent sample branches. Consequently each of the sample trees was marked with a tag in the lower crown region. All samples were subsequently obtained from immediately adjacent branches collected from the midcrown region directly above the tagged location.

Branch samples were collected individually by clipping with a polepruner fitted with a nylon catch basket. When assessing pretreatment and post treatment budworm populations each branch sample and the contents of the nylon basket were removed and placed into a cardboard "beating" box which had both ends open and which was placed on top of a white cotton drop cloth. Branch samples were processed individually by vigorously rapping the branch against the inside walls of the beating box to dislodge the larvae (or the pupae as appropriate). All larvae (pupae) were counted and the development stage determined.

---

[3]   The term block as used in this paper refers to a treatment area and does not denote a block in the true statistical sense.

It was originally intended that treatment blocks would be opened for spray application when less than 15% of the larvae were in the 2nd and 3rd instars and when 95% of new shoots had unfurled. However, because of the mixed nature of the stands and the more advanced phenology of the grand fir the criteria for larval development and foliar expansion were revised. Spray blocks were opened when approximately 20-70% of the larvae were in the 5th and 6th instars. In most spray blocks 40-50% of the population was in the 5th instar.

Pretreatment larval density samples (45 cm branch tips) were taken on the day following opening and in most cases, the spray application was made on the morning of the second day. In certain circumstances, two days were required to spray the entire treatment block. All were sprayed between June 16 and June 21.

Within one to three hours of the spray application, each tree was sampled from the designated midcrown region for deposit analysis. On the occasions when only part of a treatment block was sprayed in one day, samples were collected only from sprayed sample trees. The remainder of the samples were collected as, and when, the remainder of the block was sprayed. One (or two) 45 cm midcrown branches were cut from the predetermined sample location in each sample tree.

A total of eight twigs of new growth were cut from each branch and placed in labelled storage racks. It was occasionally necessary to cut two 45 cm sample branches in order to provide sufficient foliage. In these cases, spray deposit was assessed on four twigs from each branch. Storage racks were placed in large cardboard boxes for transport and ease of storage. Storage boxes were sealed with tape and returned as quickly as possible to the laboratory where they were stored at $40°F$ prior to microscopic analysis.

At the onset of pupation (approximately three weeks after the spray application) trees were again sampled from the same midcrown region to determine the surviving budworm population as well as final defoliation. The procedure for estimating defoliation was a slight modification of that described by Twardus [10]. Defoliation was rated for a total of twenty new, i.e. current year's, buds or shoots on each branch. The rating was based on ocular estimates of missing foliage in which each new bud or shoot was given an index rating of 1 to 6, where an index of 1 represented no defoliation and a 6 represent complete defoliation. The remaining four indices were 1 - 25%, 26 - 50%, 51 - 75% and 76 - 99%, respectively. The frequency of buds or shoots falling into each of the defoliation classes was determined and a mean value calculated for each branch. All blocks (treated and control) were sampled at approximately the same time.

Microscopic Deposit Analysis

Microscopic deposit analysis was carried out on both sides of ten randomly selected needles from each twig for a total of eighty needles per sample. One of the eight, ten-needle, samples was mounted between two clear glass microscope slides. Deposits on these needles were counted and up to 1000 droplets sized. The other seven samples were mounted between layers of clear adhesive tape and clear plastic film; deposits on these samples were only counted.

From the seventy-five sample trees in each spray block, a total of sixty grand fir/Douglas-fir trees were selected for deposit analysis. In instances where it was not possible to select all sample trees from the same species, the selection was biased as much as possible toward one species - usually grand fir.

Spray deposits on both sides of the needles were analysed under variable magnification (14 x to 60 x) using binocular microscopes equipped with calibrated eyepiece micrometers. Typically, four types of deposit were seen with the aqueous B.t. formulations; spherical or globular deposits, "hemispherical" deposits, clustered deposits or smeared deposits. When being sized, deposits were recorded in one of the above categories; however, when deposits were only being counted, no distinction was made and deposits were merely recorded as "hits".

Based on the eighty needle sample, the mean spray deposit on each sample tree was calculated and expressed as the mean number of droplets per needle.

In order to calculate the deposited droplet spectrum, mathematically calculated spread factor multipliers of 0.80 and 1.0 were applied to "hemispherical" deposits and spherical deposits, respectively [11]. The deposited droplet spectrum, as well as the $D_{v0.5}$ and $D_{n0.5}$ values were calculated for each block as a whole. No distinction was made between the deposited droplet spectum on grand fir versus Douglas-fir. Clustered and smeared deposits were not used in calculating the deposited droplet spectrum.

Analytical Procedures

An initial data analysis was attempted using an analysis of covariance (ANCOVA) model. Since only one spray was applied to each particular spray block, block was considered as a nested random effect under spray treatment. The model included the effects of pre-spray larval density and deposit density as covariates along with the effects of spray block, spray treatment and the various interactions. Post-spray larval survival or post-spray defoliation was used as the response term. As will be described later, the ANCOVA model could not be justified and relationships between the variables were investigated using chi-square analysis. Categories for the continuous variables (pre-spray larval density, deposit density and the measures of efficacy) were formed using percentiles of the corresponding distributions.

RESULTS AND DISCUSSION
Spray Deposition

The results of spray deposition assessment on grand fir and Douglas fir for each of the spray blocks (where applicable) are presented in Table 1. Variation between block means was low, ranging from 1.4 to 3.6 droplets per needle on Douglas fir and 1.0 to 3.5 droplets per needle on grand fir. Variation between trees based on the mean from each 80 needle sample ranged from 0.0 to 17.1 droplets per needle. A one hundred fold variation between the lowest and the highest deposit was observed in each treatment block. This variation is typical of eastern spruce budworm spray operations in eastern Canada. In general more than 50% of samples analysed had less than 2 droplets per needle.

TABLE 1 -- Spray Deposit Density on Grand fir and Douglas-fir

Spray block	Deposit (droplets/needle)		
	X	SD	n
Grand fir			
Dipel 6AF rep 1	2.2	2.0	49
Dipel 6AF rep 2	1.5	0.9	18
Dipel 6AF rep 3	3.6	2.6	42
Thuricide 48LV rep 1	1.4	1.5	56
Thuricide 48LV rep 2	2.2	1.5	59
Thuricide 48LV rep 3	2.0	2.8	45
Douglas-Fir			
Dipel 6AF rep 1	1.4	3.2	15
Dipel 6AF rep 2	1.9	1.5	24
Dipel 6AF rep 3	1.0	1.8	35
Thuricide 48LV rep 3	3.5	0.8	16

X - sample mean   SD - standard deviation   n - sample size

The $D_{v0.5}$ values of the deposited droplet spectra for each block, as a whole, ranged from 95 μm to 148 μm (Table 2). In many cases, these values were heavily influenced by a very small number (< 2%) of droplets greater than 200 μm in diameter. On several occasions, several large droplets up to approximately 7 mm in diameter were deposited on vehicles parked within spray blocks at the time of application. Contamination of the aircraft struts and skids with spray was a problem with the Hiller 12E Soloys used in the study and it is probable that these very large deposits were the result of drips from the contaminated aircraft. The largest droplets observed on conifer foliage were only 360 μm in diameter and may have originated from the contaminated aircraft or from poor atomization.

$D_{n0.5}$ values ranged from 43 μm to 70 μm. In nearly all cases droplets in the 0 - 80 μm size range made up 80%, or more, of the total number of spray deposits on foliage, an example of this is shown in Figure 1. These results are typical of those seen in eastern spruce budworm control operations [12, 13] and generally support the findings and discussions of other workers [14 - 17].

TABLE 2 --Characteristics of Deposited Droplet Spectra

Spray Block	$D_{v0.5}$ (μm)	$D_{n0.5}$ (μm)
Thuricide 48LV rep 1	98	43
Thuricide 48LV rep 2	108	46
Thuricide 48LV rep 3	87	43
Dipel 6AF rep 1	100	59
Dipel 6AF rep 2	95	53
Dipel 6AF rep 3	148	70

Biological Evaluation

This study supported the Meacham Pilot Project which was the subject of several other studies by the U. S. Department of Agriculture, Forest Service. Due to the mixed composition of the forest and the requirements of other researchers it was rarely possible to select the initially planned number of sixty sample trees of the same species from the seventy-five evaluation trees in each spray block. Only the results for the dominant species, grand fir, summarized in Table 3, are presented here.

One of the assumptions of analysis of variance (ANOVA) is that a plot of residual values against model estimated values shows an homogeneity of variance (a residual value being the difference between the observed value and the corresponding value predicted by the model). Unfortunately the nature of the data for both larval survival and post spray defoliation produced very skewed residual plots. Most of the plots of the residuals versus the predicted values for either the raw or transformed data revealed a nonconstant variance which increased with the level of response. When post-spray larval density or percent larval survival was the response of the model, a downward sloping band of residuals was observed. Since percent survival was set equal to 100 for sample trees on which post-spray budworm population counts were greater than pre-spray counts, the observed upper bound is quite understandable. The lower bound was formed due to the presence of many zeros as the post-spray population count.

Once the artificially created upper bound was removed and post-spray survival expressed as post-spray pupal density divided by pre-spray larval density, a fan shaped pattern of increasing variance was observed. The same pattern of skewness remained following several transformations of the response and covariate values. Although analysis of variance models are generally considered to be very robust, nonconstant variance has been shown to adversely affect some models with unequal sample sizes, in which all effects are not fixed [18]. Such was the case in this instance and when the ANCOVA produced some questionable results, the model was considered inappropriate.

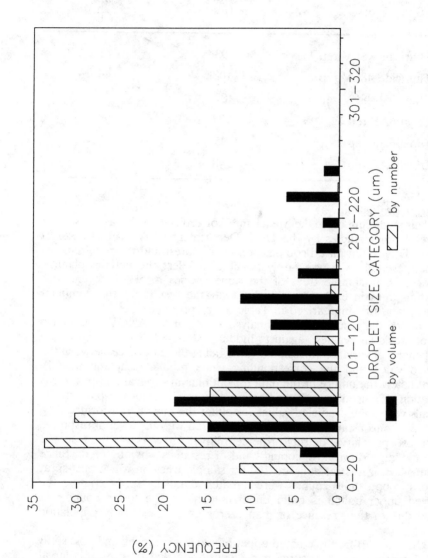

FIG. 1 -- Deposited droplet spectrum on Thuricide 48LV treatment replicate 1.

TABLE 3 -- Individual Replicate Data for the Efficacy of Dipel 6AF and Thuricide 48LV Against Western Spruce Budworm on Grand Fir

THURICIDE 48LV REPLICATE SPRAY BLOCK

	1			2			3		
	X	SD	n	X	SD	n	X	SD	n
Pretreatment larval density (larvae/branch)	20.6	11.2	56	21.0	10.7	59	20.2	13.7	45
Larval survival (%)	8.9	18.8	56	3.3	11.2	59	3.4	12.6	45
Post treatment defoliation (%)	46.9	34.5	56	47.1	23.0	59	55.3	32.6	45

DIPEL 6AF REPLICATE SPRAY BLOCK

	1			2			3		
	X	SD	n	X	SD	n	X	SD	n
Pretreatment larval density (larvae/branch)	21.9	12.0	49	22.5	11.6	18	18.3	14.2	42
Larval survival (%)	15.5	16.6	49	7.3	10.6	18	5.1	10.9	42
Post treatment defoliation (%)	25.0	22.5	49	69.0	33.3	18	40.2	26.7	42

UNTREATED CONTROL REPLICATE SPRAY BLOCK

	1			2			3		
	X	SD	n	X	SD	n	X	SD	n
Pretreatment larval density (larvae/branch)	25.3	19.9	60	18.8	12.4	60	13.2	8.6	21
Larval survival (%)	47.7	33.0	60	53.9	33.6	60	50.4	37.0	21
Post treatment defoliation (%)	52.4	37.3	60	72.8	28.4	60	60.0	27.9	21

X - sample mean   SD - standard deviation   n - sample size

Since an appropriate comparison of the treatments could not be carried out using ANCOVA it was decided to compare the treatments using chi-square analysis. Table 4 shows the results of a cross-tabulation in which post-spray population density (in effect larval survival) has been classified into two categories, zero or greater than zero.

The very significant chi-square ($p < 0.001$) shows the effectiveness of the Thuricide treatment. Additional 2 x 2 tables of this type for each spray with the control and the 2 x 2 table between the two formulations showed that Thuricide 48 LV was more effective than Dipel 6AF in killing all of the budworm. Branches from one hundred and fifteen out of one hundred and sixty trees sprayed with Thuricide 48LV had had a post-spray population density of zero. When percentage survival was classified into two or three percentile defined categories the same pattern was found (Table 5). The percentage survival in the Thuricide treatments was significantly lower than in the Dipel treatments which was significantly lower than in the unsprayed control treatments.

Inferences drawn from these results are not as strong as would have been obtained from an appropriate ANCOVA model. Cross-tabulation analysis identifies only relative differences between treatments and can not quantify the magnitude of any significant effects. i.e. provide an estimate of the absolute difference in the efficacy of the two formulations. Neither can interactions of pre-spray density or deposit density with either larval survival or post-spray defoliation be analysed. In these experiments no significant difference in pre-spray larval density was found between treatments and there was no overall significant difference in deposit density between the two spray treatments (Tables 6 and 7). Cross-tabulations of deposit density with individual spray blocks did, however, indicate some significant differences.

When percent post-spray defoliation was cross-tabulated with spray treatment, a significant chi-square was obtained (in the cases of 2 or 4 percentile defined categories for percent defoliation). In the control areas, 77/162 trees had a percent defoliation of greater than 75% (compared with 16/109 and 36/160 for the Dipel 6AF and Thuricide 48LV respectively (Table 8). In separate comparisons of the treatments, percent defoliation was again significantly higher for the control when compared to either of the treatments. The percentage defoliation was significantly lower for the Dipel than the Thuricide (for either 2 or 4 categories of percent defoliation).

Although the Thuricide 48LV produced the greatest larval mortality the percentage defoliation was least in the Dipel 6AF treatment. This is not easily explained and maybe an artifact of the sampling techniques.

At the time of spray treatment, a high proportion (20-70%) of budworm larvae were in the fifth or sixth instar and approximately 70% of the buds were unfurled. Consequently, when spray deposits were being collected, there was already noticeable defoliation on many sample branches. An evaluation of prespray defoliation in future experiments would provide a reference point from which the amount of defoliation occurring after treatment could be calculated thereby enabling a more meaningful comparision of spray treatments.

TABLE 4 -- Post-spray Population Density by Spray Treatment

Post-spray population density (pupae per branch) Frequency Percent Row Percentage Column Percentage	Treatment			
	Control	Dipel 6AF	Thuricide 48LV	Total
x = 0	156 36.19 59.54 96.30	61 14.15 23.28 55.96	45 10.44 17.18 28.13	262 60.79
x > 0	6 1.39 3.55 3.70	48 11.14 28.40 44.04	115 26.68 68.05 71.88	169 39.21
Total	162 37.59	109 25.29	160 37.12	431 100.00

Statistics for table of post-spray population density by treatment

Statistic	DF	Value	Probability
Chi-square	2	158.371	0.000
Likelihood Ratio Chi-square	2	186.268	0.000
Mantel-Haenszel Chi-square	1	156.673	0.000
Phi Coefficient		0.606	
Contingency Coefficient		0.518	
Cramer's V		0.606	

Sample Size = 431

TABLE 5 -- Larval Survival by Spray Treatment

Survival Category * (%)	Treatment			
Frequency Percent Row Percentage Column Percentage	Control	Dipel 6AF	Thuricide 48LV	Total
x < 10	17 3.94 7.39 10.49	73 16.94 31.74 66.97	140 32.48 60.87 87.50	230 53.36
10 ≤ x ≤ 38	53 12.30 56.99 32.72	27 6.26 29.03 24.77	13 3.02 13.98 8.13	93 21.58
x > 38	92 21.35 85.19 56.79	9 2.09 8.33 8.26	7 1.62 6.48 4.38	108 25.06
Total	162 37.59	109 25.29	160 37.12	431 100.00

Statistics for table of larval survival by treatment

Statistic	DF	Value	Probability
Chi-square	4	222.453	0.000
Likelihood Ratio Chi-square	4	248.627	0.000
Mantel-Haenszel Chi-square	1	191.427	0.000
Phi Coefficient		0.718	
Contingency Coefficient		0.583	
Cramer's V		0.508	

Sample Size = 431

*   Maximum row category values defined by 50th, 75th and 100th percentiles.

TABLE 6 -- Pre-spray Larval Density by Spray Treatment

Pre-spray Larval Density Category [*] (larvae/branch) Frequency Percent Row Percentage Column Percentage	Treatment			
	Control	Dipel 6AF	Thuricide 48LV	Total
x < 11	39 9.05 43.82 24.07	21 4.87 23.60 19.27	29 6.73 32.58 18.13	89 20.65
11 ≤ x < 18	45 10.44 37.50 27.78	31 7.19 25.83 28.44	44 10.21 36.67 27.50	120 27.84
18 ≤ x < 26	36 8.35 33.64 22.22	33 7.66 30.84 30.28	38 8.82 35.51 23.75	107 24.83
x ≥ 26	42 9.74 36.52 25.93	24 5.57 20.87 22.02	49 11.37 42.61 30.63	115 26.68
Total	162 37.59	109 25.29	160 37.12	431 100.00

Statistics for table of pre-spray larval density by treatment

Statistic	DF	Value	Probability
Chi-square	6	5.212	0.517
Likelihood Ratio Chi-square	6	5.135	0.527
Mantel-Haenszel Chi-square	1	1.918	0.166
Phi Coefficient		0.110	
Contingency Coefficient		0.109	
Cramer's V		0.078	

Sample Size = 431

[*]   Maximum row category values defined by 25 th, 50th, 75th and 100th percentiles.

TABLE 7 -- Spray Deposit by Spray Treatment

Deposit Category [*] (droplets/needle)	Treatment		
Frequency Percent Row Percentage Column Percentage	Dipel 6AF	Thuricide 48LV	Total
$x \leq 0.5$	20 7.43 33.33 18.45	40 14.87 66.67 25.00	60 22.30
$0.5 < x \leq 2.1$	38 14.13 36.54 34.86	66 24.54 63.46 41.25	104 38.66
$x > 2.1$	51 18.96 48.57 46.79	54 20.07 51.43 33.75	105 39.03
Total	109 40.52	160 59.48	269 100.00

Statistics for table of spray deposit by treatment

Statistic	DF	Value	Probability
Chi-square	2	4.794	0.091
Likelihood Ratio Chi-square	2	4.786	0.091
Mantel-Haenszel Chi-square	1	4.278	0.039
Phi Coefficient		0.133	
Contingency Coefficient		0.132	
Cramer's V		0.133	

Sample Size = 269

[*] Maximum row category values defined by 50th, 75th and 100th percentiles.

TABLE 8 -- Post-spray Defoliation by Spray Treatment

Post-spray Defoliation Category[*] (%) Frequency Percent Row Percentage Column Percentage	Treatment Control	Dipel 6AF	Thuricide 48LV	Total
x < 25	29 6.73 27.36 17.90	40 9.28 37.74 36.70	37 8.58 34.91 23.13	106 24.59
25 ≤ x ≤ 50	21 4.87 19.63 12.96	35 8.12 32.71 32.11	51 11.83 47.66 31.88	107 24.83
50 ≤ x < 75	35 8.12 39.33 21.60	18 4.18 20.22 16.51	36 8.35 40.45 22.50	89 20.64
x ≥ 75	77 17.87 59.69 47.53	16 3.71 12.40 14.68	36 8.35 27.91 22.50	129 29.93
Total	162 37.59	109 25.29	160 37.12	431 100.00

Statistics for table of post-spray defoliation by treatment

Statistic	DF	Value	Probability
Chi-square	6	53.728	0.000
Likelihood Ratio Chi-square	6	54.426	0.000
Mantel-Haenszel Chi-square	1	17.899	0.000
Phi Coefficient		0.353	
Contingency Coefficient		0.333	
Cramer's V		0.250	

Sample Size = 431

[*]   Maximum row category values defined by 25 th, 50th, 75th and 100th percentiles.

Unlike chemical insecticides *Bacillus thuringiensis* has to be ingested to be effective. The chances of a larva consuming a droplet of *B.t.* is increased by having a high deposit density and by post spray conditions which are conducive to active feeding [19]. Fast and Regniere have shown with the eastern spruce budworm that if a larva does not ingest a lethal dose in a short period of feeding activity and acquires a sub-lethal dose it will typically show signs of feeding inhibition and a slowing down in development. In such cases the larvae generally recover and resume feeding within several days [19].

The toxicity of currently registered *B.t.* formulations is markedly reduced by both sunlight and rainfall and the persistence or residual toxicity of formulations can be measured in days or even hours [20 - 22]. If a budworm larva does not ingest a toxic dose the first time around it is probable that by the time the larva resumes feeding, the remaining *B.t.* deposits on the foliage will have minimal residual toxicity. In this study, no precipitation or abnormally high/low temperatures were measured in the area following the first B.t. application on June 17.

Although both Thuricide 48LV and Dipel 6AF have the same nominal potency when assayed against *Trichoplusia ni* (Hubner) it is possible that the susceptibility of *C. occidentalis* differs for the two products. This, along with differences in palatability or persistance of the formulation, could explain differences in biological effectiveness between the two products.

CONCLUSIONS

Compared with untreated controls, Dipel 6AF and Thuricide 48LV gave significant reductions in larval survival and post treatment defoliation on grand fir. The Thuricide 48LV treatment resulted in significantly greater larval mortality, although a greater reduction in defoliation was observed with the Dipel 6AF formulation. Pre-spray larval densities and spray deposit densities were comparable for both treatments.

The ANCOVA model could not be justified and was considered inappropriate for this data set. This technique may be suitable for the analysis of similar field experiments with balanced designs, however, changes in population and defoliation sampling techniques may be required.

ACKNOWLEDGEMENTS

This research was funded by the USDA-Forest Service. The authors would like to thank M. C. Lambert, L. Stipe, B. Hostetler, S. Grove, A. Anderson, A. Eglitis, P. Skyler, L. Barber and J. Cota for their assistance and co-operation.

REFERENCES

[1]   Lewis, F. B., "Formulation and Application of Microbial Insecticides for Forest Insect Pest Management: Problems and Considerations," Pesticide Formulations and Application Systems: Third Symposium, ASTM STP 828, T. M. Kaneko and N. B. Akesson, Eds., American Society for Testing and Materials, Philadelphia, 1983, pp. 22-31.

[2]   Varty, I. W. and Holmes, S. E., "Heterogeneity of Spray Deposit and Efficacy within a Single Swath Applied by Aircraft over Forest Infested with Spruce Budworm, *Choristoneura fumiferana* (Clem.)," Forestry Canada Information Report M-X-168, 1988.

[3]   Van Frankenhuyzen, K. and Nystrom, C. W., "Effect of Temperature on Mortality and Recovery of Spruce Budworm (Lepidoptera:Tortricidae) Exposed to *Bacillus thuringiensis* Berliner," Canadian Entomologist, Vol. 119, 1987, pp. 941-954.

[4]   Fast, P. G., Kettela, E. G. and Wiesner, C. J., "Measurement of Foliar Deposits of *B.t.* and their relation to efficacy," in Proceedings of the Symposium:Microbial Control of Spruce Budworms and Gypsy Moths, CANUSA Spruce Budworm Program Report # GTR-NE-100, USDA-Forest Service, Broomall, PA 19008. 1985.

[5]   Armstrong, J. A. and Yule, W. N., "The Distribution of Aerially Applied Spray Deposits in Spruce Trees," Canadian Entomologist, Vol. 110, 1978, pp. 1259-1267.

[6]   Joyce, R. J. V. and Beaumont, J., "Collection of Spray Droplets and Chemical by Larvae Foliage and Ground Deposition," in Control of Pine Beauty Moth by Fenitrothion in Scotland 1978. A. V. Hoden and D. B. Bevan Eds. Forestry Commission, UK. pp. 63-80.

[7]   Payne, N., "Canopy Penetration and Deposition of small Droplets," in Proceedings of a Symposium on Aerial Application of Pesticides in Forestry, G. W. Green, Ed., National Research Council Associate

Committee on Agricultural and Forestry Aviation Technical Note 18, 29197, Ottawa, 1988, pp. 95-101.

[8]   Bryant, J. E., W. G. Yendol, and M. L. McManus, "Distribution of Deposit in an Oak Forest Following Application of *Bacillus thuringiensis*," in Proceedings of a Symposium on Aerial Application of Pesticides in Forestry, G. W. Green, Ed., National Research Council Associate Committee on Agricultural and Forestry Aviation Technical Note 18, 29197, Ottawa, 1988 pp. 261-265.

[9]   Wiesner, C. J., "Droplet Deposition and Drift in Forest Spraying", in Chemical and Biological Controls in Forestry, W. Y. Garner and J. Harvey, Eds., American Chemical Society, Washington, D. C., 1984 pp. 139-151.

[10]   Twardus, D. B., "Surveys and Sampling Methods for Population and Damage Assessment," in Managing Trees and Stands Susceptible to Western Spruce Budworm, Martha H. Brookes, J. J. Colbert, Russel G. Mitchell adn R. W. Stark, Eds. CANUSA Spruce Budworm Program Technical Bulletin #1695, USDA-Forest Service, Washington, DC. 1985.

[11]   Sundaram, A., "Drop Size Spectra, Spreading, and Adhesion and Physical Properties of Eight *Bacillus thuringiensis* Formulations Following Spray Application Under Laboratory Conditions," Pesticide Formulations and Application Systems: International Aspects 9th Volume, ASTM STP 1036, James L. Hazen and David A. Hovde, Eds., American Society for Testing and Materials, Philadelphia, 1989, pp. 129-141.

[12]   Fast, P. G., Kettela, E. G. and Wiesner, C. J., *Bacillus thuringiensis. Foliar Deposition and Efficacy Against Eastern Spruce Budworm.* Research and Productivity Council Report No. C/83/221. Fredericton, New Brunswick, Canada, 1984.

[13] Fast, P. G., Kettela, E. G. and Wiesner, C. J., Assessment of the Influence of Concentration and Foliar Deposition on the Efficacy of *Bacillus thuringiensis.* Research and Productivity Council Report No. C/85/047. Fredericton, New Brunswick, Canada, 1985.

[14] Barry, J. W., Ciesla, W. M., Tysowsky, M. and Ekblad, R. B., "Impaction of Insecticide Particles on Western Spruce Budworm Larvae and Douglas-fir Needles." Journal of Economic Entomology, Vol. 70, No. 3, 1977, pp. 387-388.

[15] Barry, J. W. and Ekblad, R. B., "Deposition of Insecticide Drops on Coniferous Foliage." Transactions of the American Society of Agricultural Engineers, Vol. 21, No. 3, 1979, pp. 438-441.

[16] Himel, C. M., "The Optimum Size for Insecticide Spray Droplets." Journal of Economic Entomology, Vol. 62, No. 4, 1969, pp. 921-925.

[17] Barry, J. W., "Deposition of Chemical and Biological Agents in Conifers", in Chemical and Biological Controls in Forestry, W. Y. Garner and J. Harvey, Eds., American Chemical Society, Washington, D. C., 1984 pp. 117-137.

[18] Montgomery, D.C., Design and Analysis of Experiments, 2nd Edition, John Willey and Sons Inc., New York, 1984, pp. 90-92.

[19] Fast, P. G. and Regniere, J., "Effect of Exposure Time to *Bacillus thuringiensis* on Mortality and Recovery of the Spruce Budworm (Lepidoptera:Tortricidae)." Canadian Entomologist. Vol. 116, 1984, pp. 123-130.

[20] Beckwith, R. C. and Stelzer, M. J., "Persistence of *Bacillus thuringiensis* in Two Formulations Applied by Helicopter Against the Western Spruce Budworm (Lepidoptera:Tortricidae) in North Central Oregon." Journal of Economic Entomology, Vol. 80, 1987, pp. 204-207.

[21] Van Frankenhuyzen, K. and Nystrom, C. W., "Residual Toxicity of a High Potency Formulation of *Bacillus thuringiensis* to Spruce Budworm (Lepideptera:Tortricidae)." <u>Journal of Economic Entomology,</u> Vol. 82, 1988, pp. 868-872.

[22] Morris, O. N., "Protection of *Bacillus thuringiensis* from Inactivation by Sunlight." <u>Canadian Entomologist,</u> Vol. 115, 1983, pp. 1215-1227.

Author Index

Subject Index

A

Adjuvants, 158, 193
Aerosol science, 10
Alachlor, 24
Application technology, 247
ASTM test methods,
 D 56: 149
 D 93: 149
 E 1116: 97
Atmospheric dispersion, 10
Atrazine, 24

B

Bacillus thuringiensis, 271
Beacon, 258
Biocontrol agents, 173

C

Closed system, 183
Compatibility, 73, 121, 134
Compression, 105
Computer codes, 10
Controlled droplet
 application (CDA), 183
Controlled release, 48, 57
Cosolvents, 149

D

Diazinon, 57
Disintegration, 105
Dispersion, 134
Drift, 193
Droplet atomization, 193
Drop spectra, 193
Dry flowable, 134

Dynamic surface tension, 158

E

Efficacy, 271
Efficiency, 183
Emulsifiable concentrates, 73
Emulsion stability, 158
Encapsulation, 33, 41, 97
Excipients, 105

F

Fertilizer compatibility, 134
Flash point, 149
Forest weeds, 247
Formulations, 73, 121

G

Granules, 48
Ground water, 41

H

Hardness, 105
Herbicides, 41
Hydrophilic:lipophilic balance
 (HLB), 258

I-J

Inert ingredients, 3
Johnsongrass, 218